BLACK
HOLES,
QUASARS,
& THE
UNIVERSE

BLACK HOLES, QUASARS, & THE UNIVERSE

HARRY L. SHIPMAN
University of Delaware

HOUGHTON MIFFLIN COMPANY BOSTON
Atlanta Dallas Geneva, Illinois Hopewell, New Jersey
Palo Alto London

Printed in the U.S.A.

Library of Congress Catalog Card Number: 75-19535

ISBN: 0-395-20615-4 (paper)
0-395-24342-4 (cloth)

CONTENTS

1 BLACK HOLES

THE UNIVERSE

In the last decade, astronomers have discovered that violent, explosive phenomena play a vital role in cosmic evolution. We see that the end of a star's life is often accompanied by violent collapse, during which the mass of the star is compacted to unbelievable densities. Some stars may end their lives as black holes, bodies so compact that light cannot escape from the space around them. We have discovered that violent explosions rock the cores of many galaxies, and that these exploding galaxies may be related to the quasars, the most distant objects that we can see. Cosmology, the study of the entire universe, has undergone a renaissance in the last decade with the increasing importance of observations to cosmological research.

Many astronomy books have neglected this new astronomy, relegating exciting topics like quasars and black holes to a few inconspicuous paragraphs. Perhaps authors hesitate to commit themselves to the printed word in such a fast-moving field, but enough is known now for the basic picture to be fairly well understood. Nonscientists who read a book like this one can fully appreciate the dynamic nature of science.

I believe that this book can fill three needs. First, those of interested non-astronomers. I have numerous friends who have read magazine articles on the topics discussed here and asked me to recommend a book for them. Having none to recommend, I have written this. Next, the need in university courses for a textbook on modern astronomy that would be suitable for nonscientists. And finally, the need for a supplemental textbook on this subject to accompany the standard textbook used in a comprehensive introductory astronomy course accommodating non–science majors or science specialists.

I thank all my astronomy mentors, particularly Owen Gingerich, Jesse Greenstein, Bev Oke, and Steve Strom, for inspiration throughout my career. Mike Zeilik, Dick McCray, Kenneth Brecher, and Remo Ruffini helped greatly by reading early drafts of this manuscript. Dave Shaffer and Mike Simon provided useful comments on Chapter 8 and Figure 9-1. The students at Pierson College, Yale University, who first suggested that I teach a course on quasars and black holes for non–science majors generated the idea for this work. Pierre Mali, Donald Richards, and the English department at Kingswood School contributed by teaching me how to write an English sentence. My wife Wendy has put up with my spending many long hours on the project. My research, which corroborates many of the ideas expressed here, is supported by the Research Corporation and the National Science Foundation (GP-44344).

PRELIMINARY: BASIC ASTRONOMICAL TERMINOLOGY

This book really starts in Chapter 1, not here. The Preliminary section contains background material that you may or may not want to read, depending on your background in astronomy and your self-confidence. I suspect that you really picked up this book because you wanted to learn about black holes and quasars, and not about parallaxes and magnitudes. However, in reading about these exciting astronomical objects, you will occasionally need basic astronomical knowledge. Since reader background will vary, I included all the basic material in this section, apart from the main body of the book. This section is intended as a reference, which you may want to peruse from time to time as you read.

The basic physics needed for astronomy, which underpins some of the forthcoming material, is here. It includes the building blocks of matter — electrons, protons, and neutrons; the relation between light and energy; the spectra of stars and their uses; and the ways astronomers and physicists measure radiation from the stars.

If you want to go on to Chapter 1, please do so. Or if you really want to read this section first, I shall not stop you.

Matter

If you divide matter — any form of matter — into smaller and smaller pieces, you discover that all matter is organized on three main levels. Sugar, for example, is composed of sugar molecules (see Figure P-1). Each sugar molecule has the properties of sugar: it is sweet, easily digested, and soluble in water, and has food value. If you try to divide a sugar molecule, you do not have sugar any more. You have atoms.

Atoms are a lower level of organization of matter. All molecules are composed of atoms, and the number and arrangement of different types of atoms within a molecule determine the properties of that molecule. Sugar contains 12 carbon, 11 oxygen, and 22 hydrogen atoms. All atoms of a particular chemical element are similar. Any of the oxygen

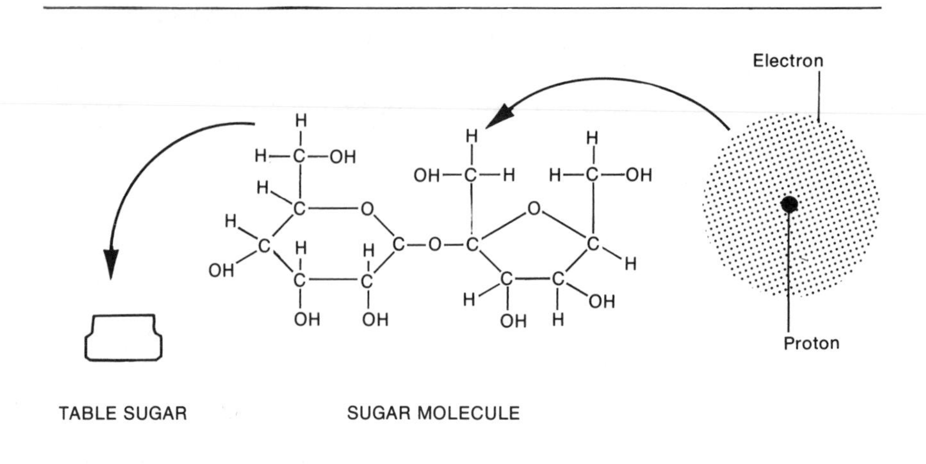

FIGURE P-1 Sugar is composed of sugar molecules, each of which possesses the properties of sugar. The molecules are composed of atoms, which in turn are composed of protons, electrons, and neutrons.

atoms of sugar could be replaced by any other oxygen atom in the universe and you would still have sugar.

If you could examine atoms closely, you could see what they are made of. Just as molecules are made of smaller units — the atoms — atoms are also made of smaller units. Any atom contains a nucleus surrounded by a number of electrons. While the electrons orbit the nucleus, they are moving so fast that they are correctly visualized as a blur around the nucleus. Figure P-2 shows a photograph of a molecule: the electron clouds are clearly visible as blurs surrounding the invisible nuclei. The nucleus, at the center of the atom, contains protons and neutrons, which are much heavier than the electrons. Protons and electrons have equal charges, which are of opposite sign, and neutrons (as their name implies) have no charge. An atom in its normal state has equal numbers of protons and electrons, so it too has no charge as a whole. The number of the electrons in an atom governs how the atom interacts with other atoms in forming molecules; thus the number of electrons in an atom determines what element that atom is.

Subatomic particles, like electrons, protons, and neutrons, can interact with each other in differing ways. The simplest interaction occurs when a neutron outside of an atomic nucleus decays into a proton and an electron. Other interactions are more complex. Collision between two protons starts a chain of reactions that makes the sun shine, the first step of which is the formation of heavy hydrogen, or deuterium, when

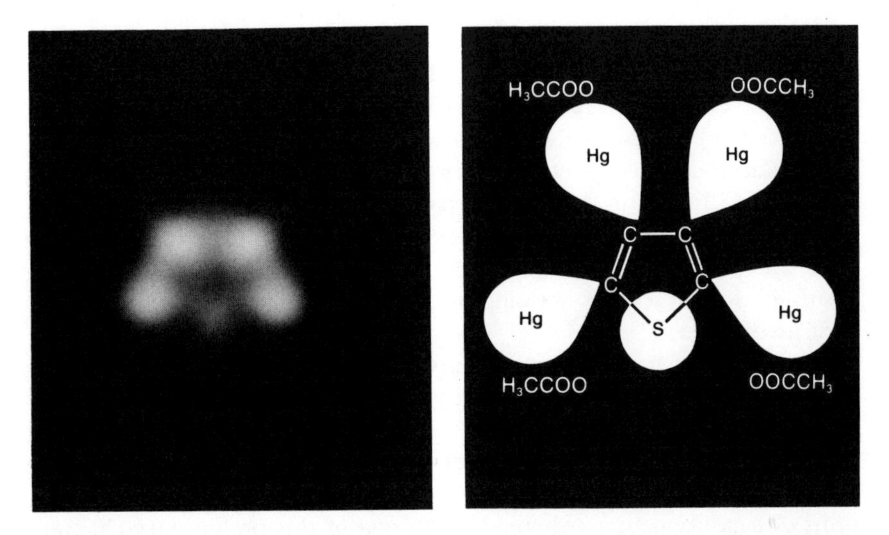

FIGURE P-2 *Left:* A photograph of a molecule of 2,3,4,5-tetraacetooxymercurithiophene. Only the mercury and sulfur atoms can be seen in this photograph; the other atoms are too small to be seen. *Right:* A drawing of the structure of this molecule, showing how the mercury and sulfur atoms are bound in the molecule. (Photographs taken from "Image of a Sulfur Atom," by F. P. Ottensmeyer, E. E. Schmidt, and A. J. Olbrecht, *Science*, vol. 179, pp. 175–176, January 12, 1973; photograph copyright 1973 by the American Association for the Advancement of Science. Courtesy *Science* and F. P. Ottensmeyer.)

one of the protons changes to a neutron. Such interactions are central to the evolution of stars and other astronomical objects.

Is there any level of organization below protons, neutrons, and electrons? This question lies at the frontier of particle physics. Most physicists studying high-energy phenomena believe that there is a way that a proton, for example, can be taken apart to reveal its building blocks; these have been named quarks. However, since no one has yet found a quark, the theory remains unproven.

Light and energy

Almost all of our information about astronomy comes to us through light and other forms of electromagnetic radiation such as radio waves, x-rays, and infrared radiation. No one has yet brought us a piece of a star, although pieces of the moon have made their way to earth via the Apollo program. Since light is the source of virtually all of our knowledge of astronomy, its nature is fundamental to any study of the subject.

Light is a form of energy. It is difficult to define precisely the term *energy,* much in the news recently. It is easier to develop a mental picture of the energy concept by asking what energy does. The usual definition of *energy* is "the ability to do work," which is an accurate description. The type of work that energy does varies with its form. Light energy can illuminate the printed page; heat energy can keep us warm; kinetic energy (or energy of motion) can be used to move something from one place to another. All activity can be thought of as involving manipulation of energy – changing the character of energy from one form to another.

Whatever happens in the universe can neither create nor destroy energy. Thus a fundamental question about any astronomical phenomenon is, What is the energy source? The sun bathes the earth in energy, as its rays deposit 1,388,000 ergs of energy per second on each square centimeter of surface area on top of the atmosphere. (Ergs and all other units of measurement are defined at the end of this preliminary chapter.) This energy comes from nuclear reactions taking place in the center of the sun, and is transformed from nuclear energy into light energy by the solar furnace. We intercept and use some of this energy, changing it to different forms. But whatever happens, energy is neither created nor destroyed. This principle is so important that it can be called a physical law, the First Law of Thermodynamics: Energy is neither created nor destroyed.

The energy contained in light comes in little packets. Each packet, a photon, carries a certain amount of energy. The amount of energy depends on the character of the photon; an energetic x-ray photon has trillions of times as much energy as a low-energy radio photon.

Light is a form of electromagnetic radiation. Electromagnetic radiation can be understood most easily, perhaps, by thinking of radio waves. An electron in a radio or a TV antenna feels forces pushing it along the antenna, first one way and then the other. These forces come from the electric and magnetic fields carried by the radio wave. All radiation contains these electric and magnetic fields, reversing direction at periodic intervals.

The forms of electromagnetic radiation differ in how quickly the reversal of direction takes place. Reversals come much less often in a radio wave than in a light wave. The rate at which these reversals take place determines the frequency of the wave. The frequency equals the number of times the electric and magnetic fields go through complete cycles in one second. Frequency is measured in hertz, or cycles per second. For example, if your radio is tuned to 1100 kilohertz, in the middle of the AM dial, the radio is picking up the particular electromagnetic waves in which the fields reverse direction 1,100,000 times per second.

Closely related to frequency is wavelength. If you measure the distance that a light wave travels in one cycle of the change in field direction, that distance is one wavelength. If the frequency of the wave is high, the electric field reverses direction rapidly and the wave does not have to travel far for the field to go through a complete cycle. Thus high-frequency waves have short wavelength and low-frequency waves have long wavelength.

Electromagnetic radiation comes in all varieties, from very energetic gamma rays of very short wavelength and high frequency on one end of the spectrum to radio waves of long wavelength, low frequency, and low energy on the other. Historically, several different names have emerged to characterize different parts of the spectrum, and these names are listed in Figure P-3, along with wavelengths, frequencies, and energies of typical photons. (Powers-of-ten notation, used in this figure, is explained later in the section.) The scale at the bottom of the spectrum illustrates what a small part of the spectrum visible light is. Below the scale are examples of the kinds of instruments used to pick up the different types of radiation. The atmosphere blocks a great part of the electromagnetic spectrum from our view, since only visible light and radio waves can travel unimpeded to the ground. For some parts of the spectrum, like the infrared, radiation is blocked by water vapor in the lower part of the atmosphere. To see this radiation you must fly above the lower atmosphere in an airplane or balloon. (There are a few infrared frequencies that are not blocked by water vapor.) For other parts of the spectrum, such as the ultraviolet, satellites are necessary to see what objects in space are doing. (See Figure P-4.)

The astronomer is interested in the great extent of the electromagnetic spectrum because different types of objects emit different types of radiation. There are several ways in which light can be emitted from an object. Here we consider the most common – radiation from hot objects.

Thermal radiation is the usual term for characterizing light emission from hot objects. Anything that is hot radiates, generally producing photons with roughly the same amount of energy that is present in the individually moving particles. Heat comes from the motion of molecules in an object, and the temperature of the object indicates how much energy each molecule in the object possesses. Consider a particular part of the electromagnetic spectrum, red light. That type of radiation will be emitted by objects whose moving molecules have the same energy as a photon of red light. Such objects are at temperatures of 5900 Kelvins, which happens to be the temperature of the sun. Infrared photons, lower-energy photons, are emitted by cooler objects. The right-hand column in Figure P-3 lists the temperatures of objects that emit the types of radiation listed. (Objects emit radiation over a wide range of wave-

RADIATION TYPE	WAVELENGTH (METERS)	FREQUENCY (HERTZ)	PHOTON ENERGY (ELECTRON-VOLTS)	CORRESPONDING TEMPERATURE (KELVINS)
GAMMA RAYS	1.2×10^{-13}	2.5×10^{21}	10^{7}	3×10^{11}
X-RAYS	1.2×10^{-9}	2.5×10^{17}	10^{4}	3×10^{7}
ULTRAVIOLET	1.5×10^{-7} (1500 Å)	2×10^{15}	8.3	24,500
BLUE	4×10^{-7} (4000 Å)	7.5×10^{14}	3.1	9150
RED	6×10^{-7} (6000 Å)	5×10^{14}	2.0	5900
INFRARED	2×10^{-6}	1.5×10^{14}	0.62	1850
MICROWAVE	3×10^{-4}	10^{12}	0.004	9.3
RADIO	1	3×10^{8}	1.2×10^{-6}	0.0035

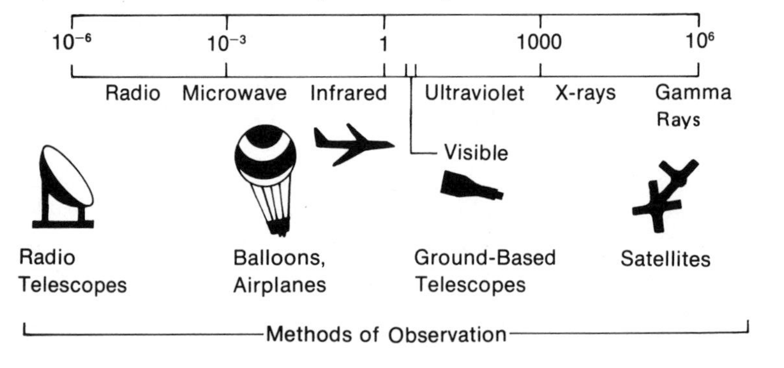

FIGURE P-3 Different types of radiation (see text).

lengths. The temperature listed in the figure is the temperature of an object that emits the maximum number of its photons at the wavelength indicated. Furthermore, since the characteristics of an object can influence the type of radiation emitted, the temperatures listed are for objects without intrinsic color, namely blackbodies.)

The last column of the figure provides some insight into the types of objects that are likely to be found by astronomers studying various parts of the spectrum. Stars, for example, have temperatures roughly between 1000 and 100,000 K, thus they will be found by astronomers studying the infrared, optical, and ultraviolet regions. X-ray astronomers will see hotter objects, such as streams of gas falling into a black hole, heated up by the hole's gravity before the gas falls in. Astronomers studying the microwave region will pick up very cool objects, such as the radiation left over from the Big Bang at a temperature of a few degrees. There are no objects in the natural universe with temperatures of 0.0035 K, so the radio astronomer does not pick up thermal radiation. Instead, he picks up nonthermal radiation, which in quasars comes from fast electrons spiraling in a magnetic field (see Chapter 8).

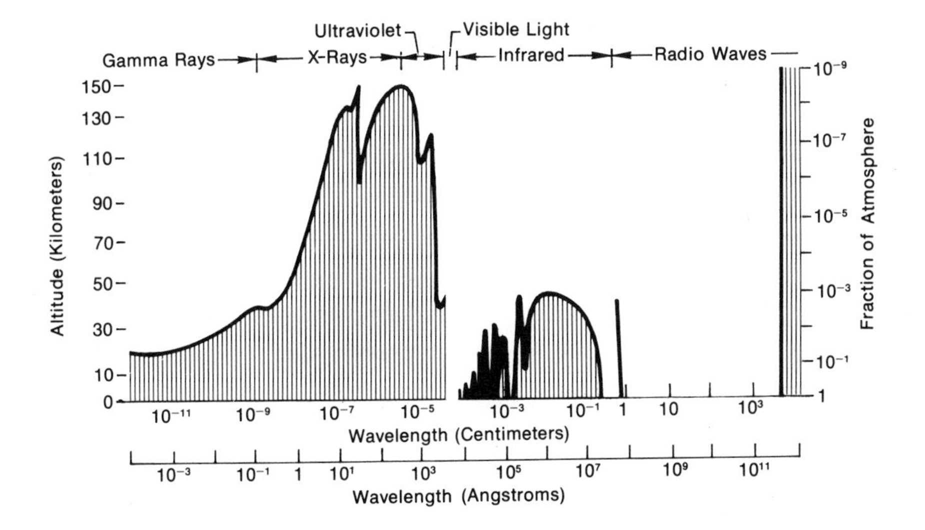

FIGURE P-4 Absorption of radiation in the atmosphere. The upper boundary of the dark area specifies the altitude where the intensity of celestial radiation is cut down to half its original value by the atmosphere. Ultraviolet and x-ray observations must be made from high-altitude rockets or satellites; infrared observations require high-altitude mountain observatories in dry climates or airplanes. (From Leo Goldberg, "Ultraviolet Astronomy," *Scientific American*, June 1969. Copyright © 1969 by Scientific American, Inc. All rights reserved.)

Stars and their properties

Before 1950, astronomers obtained virtually all their information about the universe from visible light, a tiny fragment of the electromagnetic spectrum. Most visible light in the universe is emitted by stars, and astronomy had so far concerned itself primarily with stars. The extension of our view to other parts of the electromagnetic spectrum has introduced astronomers to other types of objects, such as quasars and perhaps black holes. Stars do come into the picture in a supporting role, and a brief review of the properties of stars is thus in order.

A star is a ball of gas that produces light. One question to ask about a star is exactly how much light it is producing. An answer involves measuring the star's brightness as seen from the earth and the star's distance from the earth. These are separate tasks.

STELLAR MAGNITUDES

Deferring to historical custom, astronomers measure the apparent brightness of stars through the magnitude scale, a very confusing way. This scale originated when Hipparchus made the first star catalogue in the second century B.C. He classified the stars according to their brightness, labeling the brightest stars first magnitude, the next brightest second magnitude, and so on down to fifth magnitude. Since his catalogue was so useful, astronomers kept the magnitude scale rather than adopt a more logical one. Much confusion about the magnitude scale can be alleviated if you remember that it goes backwards: big (positive) magnitudes mean faint stars.

The magnitude scale has been made more quantitative since the days of Hipparchus, but the main features are still with us. Magnitudes can be measured to two decimal places by putting a photometer (a device that acts essentially like the automatic light meter in a camera) at the back end of a telescope and seeing how much light is hitting the photometer. Stars that differ in brightness a hundredfold differ in magnitude by 5. (Thus the magnitude m equals -2.5 log (amount of light) plus a constant.)

STELLAR DISTANCES

Distances of stars must be known if one is to determine whether a particular bright star is a big one far away or simply a nearby one. For example Rigel, in Orion's foot, and Sirius are both very bright stars in the winter sky. (Figure 2-2 shows where to find these stars.) Sirius is quite close, while Rigel is much farther away. If these two stars were at the same distance from the earth, Rigel would be much brighter.

The distances to nearby stars are measured by triangulation, or parallax (see Figure P-5). The astronomer examines the position of the star from opposite ends of the earth's orbit. If the star is close, it will subtend a large angle as viewed from the two ends of the orbit. If it is far away, the angle will be smaller. The size of the angle thus determines the distance of the star. Now you can tell whether it is a bright star that is close or a still brighter star far away.

The intrinsic brightness of a star, or how much energy it puts out into space, is the star's luminosity. To measure this, you measure the star's magnitude as seen in the sky, or its apparent magnitude, and its distance from you. If you know that distance, you can then calculate how bright it would be if it were some standard distance, such as ten parsecs, from the earth. A star's magnitude at a distance of ten parsecs from the earth is its absolute magnitude.

Absolute magnitude is related to luminosity. For example, if a star has an absolute magnitude of 9.72, it would be five magnitudes (or a hundred times) fainter than the sun would be at that distance. As the sun has a luminosity of 3.9×10^{33} ergs/sec, this star would have a luminosity smaller by a hundred times, or 3.9×10^{31} ergs/sec.

STELLAR TEMPERATURES AND SPECTRA

Once you know the luminosity of a star, a relevant question is whether it is luminous because it is large but cool or small and hot. You want its surface temperature. One way to determine a star's temperature

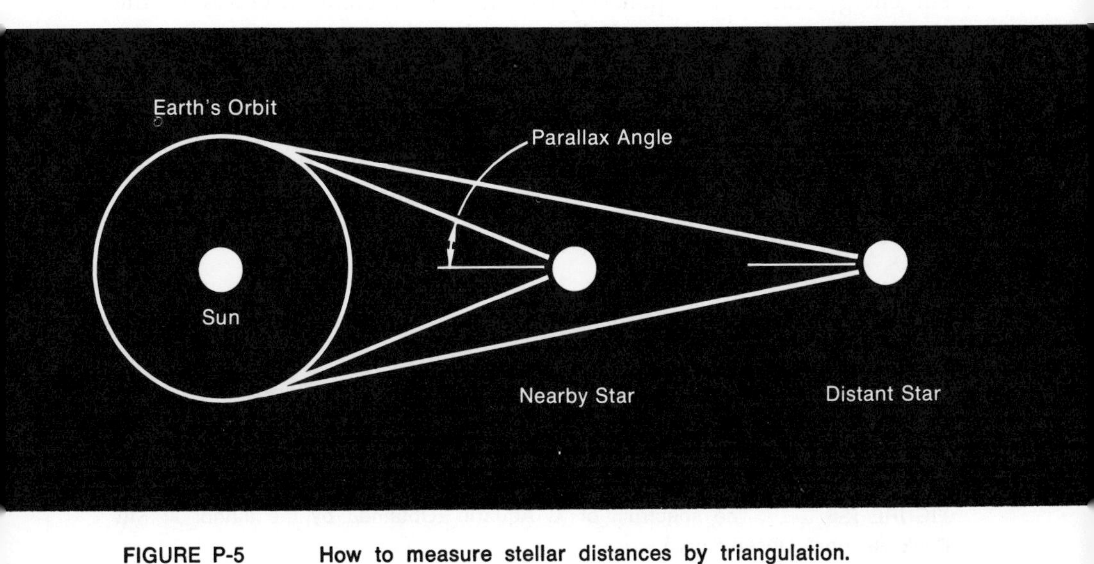

FIGURE P-5 How to measure stellar distances by triangulation.

is to look at its color; hot stars are blue and cool stars are red. The electromagnetic spectrum figure, Figure P-3, shows this result. Another way to determine the temperature of a star is to look at its spectrum.

If you spread the light of a star out into a rainbow and photograph it, you have a spectrum of the star. A spectrum of Xi Aquarii is shown in Figure P-6, with the colors going from blue to red. The spectrum is just an image of the star in various wavelengths. The spectrum is spread out or widened from top to bottom so that it can be analyzed more easily.

The most interesting feature of the spectrum is the series of dark vertical lines crossing it. Wherever there is such a dark line, it means that the star is not emitting much light at that particular wavelength. These dark lines hold the secret of spectrum analysis, for it is through examining and interpreting these lines that we can determine the temperature and luminosity of the star. To see how these dark lines are related to the star's temperature, we need to consider where they come from. Let us consider a particular dark line, in the middle of the spectrum with a wavelength of 4340 angstroms (1 angstrom, which is abbreviated Å, equals 10^{-8} centimeter).

As radiation travels from the surface of a star toward space, it must travel through the obstacle course of atoms in the star's outer envelope. This obstacle course is filled with thief atoms, which steal energy from the radiation. These thieves are selective in what they steal. Let's follow the paths of three photons through the outer envelope, with wavelengths 4340, 4330, and 4350 angstroms – one at line center, and two on either side of the line. These photons pass by hydrogen atoms (Figure P-7). These atoms are the thieves. They can exist in a number of different energy states, corresponding roughly to different distances of the electron from the nucleus. It happens that the amount of energy the 4340-angstrom photon carries is exactly the right amount necessary to kick a hydrogen atom from its first excited state into a higher energy state, the fourth excited state. A hydrogen atom in the first excited state acts like a thief, stealing the energy of the photon so it can move up to a higher energy state. The hydrogen atom has little use for the 4330- or 4350-angstrom photons, since they carry either too much energy or too

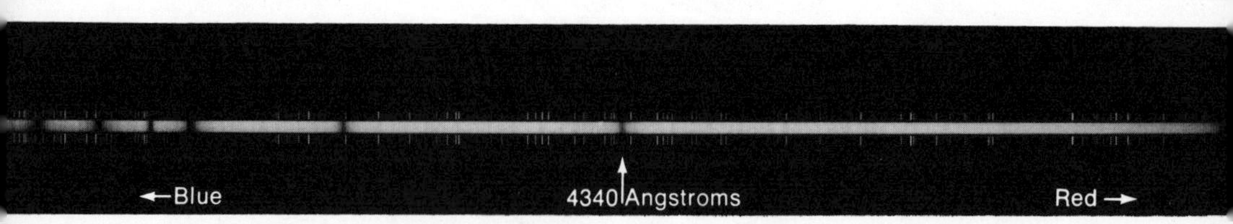

←Blue 4340 Angstroms Red →

FIGURE P-6 The spectrum of Xi Aquarii. (Obtained by the author at Kitt Peak National Observatory.)

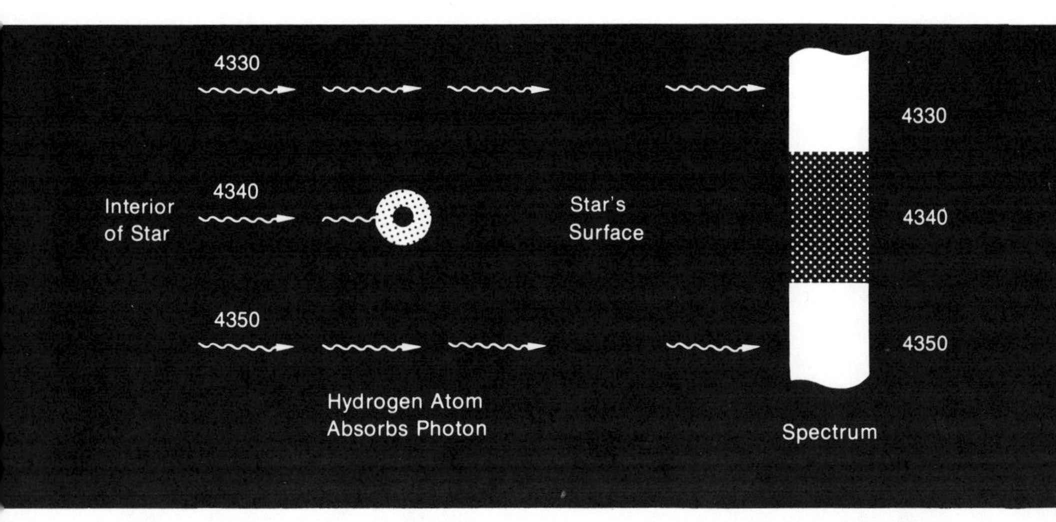

FIGURE P-7 Formation of a dark line in a stellar spectrum. Radiation flows from left to right, from the stellar interior through the surface to the space outside of the star. The 4340-angstrom photons are absorbed by hydrogen atoms, leaving a gap in the spectrum, called an absorption line.

little to move the atom to a higher energy state. Thus the obstacle course of the star's outer envelope has highly selective thieves in it, who steal the energy from only one type of photon. Some photons can travel through easily, whereas others have trouble. The photons, like the 4340-angstrom one, that are absorbed by the outer envelope create gaps, or dark lines, in the star's spectrum.

Whether a photon can make it through the obstacle course or not depends on the types of atoms in the star's outer envelope. The 4340-angstrom photons can be absorbed only by a hydrogen atom in the first excited state. If the star is cool enough that there are relatively few hydrogen atoms sitting around in this excited state, there are not enough available thieves waiting to steal these photons and there will be no gap or line in the spectrum. If the star is somewhat hotter than the sun, there are many hydrogen atoms in the first excited state and the gap will be very large, as it is in the spectrum of Xi Aquarii in Figure P-6.

Thus you can determine the temperature of a star by examining the strength or weakness of certain spectral lines such as the 4340 line (called $H\gamma$, or H-gamma). Spectral classification is now a well-developed art; someone good at this work can just look at the spectrum of a star and obtain a fairly good idea of its temperature, its luminosity, and any peculiarities it might possess. Spectra are grouped into different classes, denoted by letters of the alphabet. Historical reasons dictate that the order from hottest to coldest is not straightforward: O, B, A, F, G, K, M.

Units of measurement

Many aspects of astronomy involve measuring physical quantities and creating theoretical models to explain these measurements. The measurements are quantitative, expressed in numbers that must refer to some kind of unit.

LENGTH

Most units of length are familiar, but astronomical distances are so large that special units are often used. Distances between stars are measured in *parsecs*. A star one parsec away has a parallax angle (Figure P-5) of one second of arc, and is 206,265 times farther from the earth than the sun is. One parsec is 3.085×10^{13} km. Parsecs are convenient units for measuring stellar distances because stars are roughly one parsec apart in our part of our galaxy. One parsec is a very long distance; it is difficult to visualize exactly how long. People try analogies: if the earth were 6 inches from the sun, the nearest star would be 28 miles (1.33 parsecs) away, but somehow these analogies do not convey a true impression of how vast and empty the universe is.

ENERGY

Probably the most familiar unit of energy is the kilowatt-hour, a feature of your electric bill. Ten 100-watt light bulbs shining for an hour use up one kilowatt-hour's worth of energy. Most of the energies quoted in this book will be in *ergs*. One erg is not much energy; a kilowatt-hour is 3.6×10^{13} (36,000,000,000,000) ergs. The energy of motion (or kinetic energy) of a two-gram insect crawling along at a speed of one centimeter per second is one erg. (In English units, the bug weighs 1/14 ounce and covers an inch in 2.5 seconds.) *Electron-volts* are also used to measure energy; one electron-volt is 1.60207×10^{-12} erg. Photons of visible light, with energies of a few electron-volts, carry very little energy in terrestrial terms.

TEMPERATURE

All temperatures in this book will be expressed in Kelvins or degrees Celsius (Centigrade) above absolute zero.

ANGLES

All positions on the sky are measured in angles, because you can only measure the angle between light rays coming from different objects. Angles are measured in degrees, minutes, and seconds of arc, where 360 degrees equal one circle, 60 minutes equal one degree, and 60 seconds equal one minute. The moon and sun are half a degree across in the sky. Telescopes allow astronomers doing high-precision work to measure stellar positions with an estimated error of 0.01 second of arc, which equals the diameter of a dime 200 miles away. Such precision can be achieved only by repeating the measurement many times.

POWERS-OF-TEN NOTATION

Astronomers and also other scientists often have to deal with very large or very small numbers. To save effort, a shorthand system has been developed. A number like 3,000,000 is expressed as 3×10^6. Where does this come from? The form of a number is $X.XXX \times 10^n$, where n is the number of zeros to be added to the number, or the number of places the decimal should be shifted to the right. If the exponent is negative, the decimal point should be shifted to the left. Thus 3×10^3 is 3000; 6×10^{-2} is 0.06; and 3×10^{12} is 3,000,000,000,000. (Be sure you can figure these out.) If you prefer to think in words, the following equivalents are useful:

10^3 = thousand
10^6 = million
10^9 = billion

Further, you can extend units of measurement by using prefixes. The metric system uses these extensions a great deal, in common with measurements in astronomy. In this book I shall often talk about kiloparsecs (1 kiloparsec = 1000 parsecs) or megaparsecs (1 megaparsec = 1 million parsecs). If you like large numbers, you can use *giga-* (10^9) and *tera-* (10^{12}), but at this point we usually use powers-of-ten notation. Other prefixes are *centi-* (10^{-2}), *milli-* (10^{-3}), *micro-* (10^{-6}), and *nano-* (10^{-9}). Thus the gross national product of the United States can be thought of as exceeding one trillion dollars, one teradollar, or 10^{12} dollars. If you really want to impress people, tell them about googols and googolplexes. A googol is 10^{100}, and a googolplex is 10^{googol}, that is, $10^{10^{100}}$!

More information about the terminology of astronomy can be found in the Glossary or in an elementary astronomy textbook. "Suggestions for Further Reading" at the end of the book lists a few appropriate selections.

INTRODUCTION: THE VIOLENT UNIVERSE

Quasars and black holes are two inhabitants of the new, violent universe that the astronomy of the 1960s has revealed. Beginning with the discovery of quasars in 1963, observations have revealed that explosions and catastrophic collapse are important stages in the evolution of some astronomical objects. These explosions accelerate electrons to speeds close to the speed of light; some galaxies release as much energy in one second as our sun releases in ten thousand years. The discovery of this violent universe has changed astronomy, producing what several people in the field have called a golden age. The pace has quickened: new discoveries and new interpretations appear every month—sometimes even weekly. Each new advance poses still more questions, sometimes making the universe more puzzling, not less. We do have some ideas of what quasars and black holes are, though important details are still unclear. What we know about them makes them the most exotic and exciting objects in the universe. What we still have to learn about them could lead to a revolution in physics.

A black hole represents the ultimate triumph of gravity in its role as the regulator of a star's life cycle. When a massive star dies, gravity becomes so strong that the star cannot possibly hold itself up. What happens next? We do not know for certain. One possible end to a massive star's life is total collapse. Its tombstone is a black hole: outside, we can still feel the gravity from the hole, but we can't see it because gravity prevents light from escaping from the star's surface.

If we can't see a black hole, how do we tell what it looks like? The pencil and paper calculations of the theoretical physicist help us here. We know (or think we know) how gravity works, and we can use this knowledge to give us a good idea of what a black hole does to its surroundings. Holes provide explorations of the outer limits of Einstein's theory of gravity.

Evidentially, there are some reasons to believe that black holes are more than the products of the very fertile imaginations of theoretical physicists. X-ray astronomers, using the *Uhuru* satellite, have observed radiation that probably comes from heated gas being sucked into a black hole. Knowledge of the life cycles of stars has greatly advanced in the last decade, and black holes have appeared a likely end to the lives of massive stars.

While black holes contain the strongest gravitational forces in the known universe, quasars are probably the most energetic objects in the universe. The astronomer who examines a photograph of a quasar sees a small, starlike dot, indistinguishable at first from an ordinary star. Closer analysis of the light from quasars reveals that in all cases, this light is much redder at the earth than it was when it left the quasar. This redshift, most people believe, means that the quasars are very far away, more distant than the most distant galaxies we can see. Thus the quasars, whose central energy sources are not much bigger than the solar system in some cases, must be emitting more light than the average galaxy. This cosmological interpretation of quasars is supported by recent discoveries of galaxies that are exploding almost as violently as the quasars. The cosmological interpretation raises the question: Where does all the energy come from?

A few astronomers find the energy requirements of quasars so difficult to understand that they feel that the quasars are much closer than their redshifts indicate. These astronomers believe that quasars are companions of bright, nearby galaxies, and they present some hotly disputed photographic evidence to support their belief. If these astronomers are correct in their interpretation, then these redshifts become quite mysterious. Some factor, now unknown to physics, is causing the light from quasars to be much redder than it was when it left the quasars. If these astronomers are wrong and the conventional view prevails, then quasars are the most distant objects in the universe. Either way, it's an exciting field of research.

If quasars are very distant, they may prove the keystone in our developing a better picture of what the universe looks like and how it evolves. Cosmology, the study of the nature and evolution of the universe, has developed greatly in the last decade, stimulated by the discovery of quasars and more intensive investigation of galaxies. In 1965, physicists from Bell Laboratories discovered radio waves left over from the Big Bang, the explosion that marked the beginning of the universe. Astronomers are now seeking to determine the ultimate fate of the universe. Will it go on expanding forever or will the expansion slow down, stop, and turn into a contraction? In cosmology, as in the study of black holes, we are exploring the outer limits of Einstein's theory of gravitation. In some respects there is an analogy between the mathematical description of a black hole and the description of the entire universe.

The study of black holes, quasars, and cosmology is part of the golden age of astronomy. A definitive picture of these objects is not immediately available, for the violence in the astronomical universe is somewhat mirrored by turbulence in the astronomical community. The astronomical controversies are interesting in themselves nevertheless, for the story of black holes and quasars illuminates the methods that astronomers use when trying to interpret the universe. I hope incidentally to communicate some of the flavor, excitement, and frustration of this scientific undertaking as I describe the violent universe.

Controversies and changing interpretations of nature are common in the scientific world, despite the popular image of science as some sort of Delphic oracle. Unless you read the research literature, which is published in highly technical, condensed journal articles, most of your scientific knowledge will probably come from textbooks. Textbooks usually seek to encapsulate an area of knowledge in a neat, shiny package; as a result, the textbook reader often gains the impression that the research that led to the knowledge was equally neat and shiny. It wasn't. A researcher must find his way through a tangled maze of uncertain observations and contradictory interpretations to try to fill in the total picture.

The scientific process

Two basic conflicts underlie the so-called scientific method, or the way that scientific research is done. In our immediate story, these conflicts provide a background for the study of black holes and quasars. The first conflict is between theory and observation. It involves the attempt to match a mental model of the natural world with the evidence our senses gather. The second conflict is between the desire to uphold the currently accepted basic physical laws and the need to change them by a scientific revolution when conflict between theory and observation becomes overpowering.

MODELS AND REALITY

J. L. Synge, in *Talking about Relativity,** has described the conflict between theory and observation as an interplay between two worlds – the M-world, or model world (as I shall call it), and the R-world, or real world. The real world is the world we live in: it contains apples, oranges, stars, black holes (maybe), and other such things. Scientific enterprise seeks to understand the real world by matching it to a model world.

* Bibliographical details for works mentioned in the text can be found in "Suggestions for Further Reading" at the end of the book.

The model world exists only in people's minds. It contains castles in the air, created from the basic laws of science and mathematics. Also needed is some basic picture of what some object is like to build a model. For example, consider a model of the sun. You start with the relevant laws of physics, or prescriptions of the behavior of matter under certain conditions. You then add the basic picture: the sun is a ball of 1.989×10^{33} g of gas, which is 70 percent hydrogen, 27 percent helium, and 3 percent heavier elements. Using mathematics and often a computer, you then discover certain properties of the object you are trying to model. The computer will produce a value for the solar luminosity, for example, from the model. The real world now enters, for the model's luminosity should be the same as the sun's real luminosity, within the limits of error.

The connection between the model world and the real world comes from observations, but this connection is not always completely straightforward. You should remember that the world of the observer is a kind of model world by itself. A dot formed by the blackening of a few grains of silver bromide on a photographic plate is not a quasar; it is only an image of a quasar. You then need to interpret the image. As controversies about interpreting photographs that seem to depict luminous bridges connecting quasars and bright, nearby galaxies demonstrate, interpretation of an image is sometimes uncertain.

The matching and modeling process also takes place in other disciplines, some nonscientific. Science is unique in that the matching is quantitative. The model produces some numbers that can then be compared with observed ones. Sometimes these numbers are not completely certain, since experimental error does exist; and models themselves have some uncertainty because of the approximations made in computing them.

Matching theory to observation is sometimes called prediction; that is, a theorist tells an observer what he should see and the observer then goes out and sees if the theorist's idea is true. Some people argue that matching real and model worlds *always* involves prediction, and that prediction is the essence of science. To me, the word *prediction* is too narrow a term, especially in astronomy. We astronomers can only observe the universe; we can't do experiments with stars. As a result, theorists spend a great deal of time trying to model or explain existing patterns in the real world, predicting in reverse. I believe that a more appropriate word is *understanding*. When a model matches reality, we think we know what causes the real world to behave the way it does. (For a fuller discussion about prediction in science, see Stephen Toulmin's *Foresight and Understanding*.)

The progress of science involves both theory and observation. It is often possible for a theorist to forget that he is only a model-builder. The models become so fascinating that they become real, like the statue of Pygmalion. Synge, a theorist himself, calls fascination with models the Pygmalion syndrome. You get caught up in your own work and become oblivious to the fact that you're only dealing with pencil marks on paper (or ten-foot-high piles of computer output) and not real stars. We astronomers have a particular difficulty here because we can't touch a star. All we know is what our telescopes reveal. As a result, the astronomer can develop great faith in a theoretical picture of, say, a quasar, which has no basis in reality. You have to keep talking to your observational colleagues or go and make some observations yourself to remind yourself that you are ultimately trying to prove something about the real world; this means not simply indulging yourself in an elaborate mathematical exercise thinking it represents the real world. The Pygmalion syndrome is moderately prevalent among black hole theorists, so watch out for it. There is nothing wrong with mathematical speculation so long as you recognize it as such.

SCIENTIFIC REVOLUTIONS

But how does science advance? How are obsolete models replaced by new ones? There are always some anomalies, some areas in which the model and observations do not quite match. Sometimes all you have to do is modify the model world in a very small way. You build a new room on the castle in the air to make it look more like the real world. Less often, the change in the model must be more fundamental, since the basic laws of science, the foundations of the model, are at fault. When these laws are replaced, a scientific revolution occurs. Thomas Kuhn, in *The Structure of Scientific Revolutions,* has given a fairly complete and instructive model of what happens during a scientific revolution. I think that his model applies to the revolutions that may take place as a result of the discoveries in astronomy that this book recounts.

We seek a scheme to distinguish between big changes, or scientific revolutions, and minor ones. A big change involves a change in the foundations of the model world, or the paradigm. The paradigm, a crucial concept in Kuhn's scheme, is a set of rules in its narrowest sense. These rules are the physical laws, or the foundations of all theoretical models. "You cannot exceed the speed of light." "Energy is neither created nor destroyed."

But a paradigm is more than just a set of rules written down in textbooks. Behind the rules lies a philosophy or world view. During the Copernican revolution, the essence of the paradigm change was the realization that the earth, the home of man, was not the center of the universe. No one except the astronomers really objected to changing the rules of planetary motion. Only when Galileo made nonscientists, particularly theologians, realize that changing the rules also meant removing the earth from its central location did the hierarchy of the Catholic church realize the magnitude of the paradigm change.

Most of the time — when science is not undergoing a revolution — the processes of normal science are at work. Theoreticians are busy building and refining the castles in the air of the model world, using the existing paradigm, as they seek to match model and observation. Observers approach the matching process from the other end, discovering new patterns that need explaining and testing the models. It is like a picture puzzle, in a way. Observations tell you, for instance, that the image of Mars is always accompanied by two little dots. You expand your model of the solar system to include two moons of Mars; you name them Phobos and Deimos. Nothing basic has changed. You are still using the same foundation, the same paradigm. You have added a new room to Castle Solar System, and filled in a piece of the puzzle. There is always work of this sort to be done, as the world of science always contains a few anomalies, where theory and observation do not quite agree.

Occasionally the processes of normal science, of eliminating anomalies by building more and better models and obtaining more observations, do not work so easily. Sometimes it is impossible to build a model, based on the existing paradigm, that can explain a bothersome observation. Often the situation is not clearly drawn; you can stretch the model to explain the observations but you may be stretching it too far. When you take a reasonably simple model, a small, neat castle in the air, and patch on extra rooms and staircases and cupolas and porticos, the whole structure can begin to look awful. The paradigm becomes badly distorted by this excess architecture. Maybe the paradigm is wrong and needs to be changed. The principle that a scientific explanation must be neat, involving a minimum number of arbitrary assumptions, has proved useful in the past. It has been given the name "Occam's Razor," after a fourteenth-century philosopher.

Paradigm change completes the picture of a scientific revolution. To review: An anomaly develops and grows, as model and reality refuse to be brought together. When the anomaly has resisted repeated attack, a crisis develops. The paradigm must be changed, since the foundations are at fault. A chaotic period follows, during which a new paradigm is sought. When a new one is found and accepted, the revolution is complete and the new order takes over. The new paradigm must be able to explain the anomaly that caused all the trouble in the first place and it must be consistent with all other known facts.

Black holes and quasars: a scientific revolution?

A conflict between old and new paradigms may arise from the differing interpretations of quasar redshifts. The cosmological interpretation – the idea that expansion of the universe causes the redshifts of quasars – is adjudged by most astronomers a good, solid foundation. Most of Part Two of this book describes the satisfactory quasar model that is based on the cosmological paradigm. While there are some unanswered questions, the conventional point of view regards them interesting frontiers of research rather than potential problems with the structure. To these astronomers, a problem like the quasar energy problem will be solved by building another room, enhancing the beauty of Castle Quasar, increasing our understanding of the physical world.

A small but growing number of astronomers are becoming increasingly uncomfortable with the cosmological redshift paradigm for a number of reasons. To them, the extra room that must be added to Castle Quasar is not a beautiful new wing but a silly cupola that undermines the integrity of the whole structure. Ideally, a direct confrontation of model with reality should settle the issue, but as we cannot directly observe the heart of a quasar, its energy source, we have to use judgment to settle the issues. The challengers must face the problem of finding a new foundation, a new paradigm to explain the riddle of the redshift. No known physical mechanism can provide such an explanation except expansion of the universe.

A similar conflict may lie ahead in the black hole field. Black holes are puzzling phenomena, but they are almost entirely theoretical at this point. If they exist, they may indicate a serious flaw in Einstein's theory of gravity.

What sort of a world do we live in? Are black holes and quasars extra rooms now being added to the astronomical castle? If so, they will lead to more and more rooms, since very unusual events occur in them. Understanding their properties will enable us to explore much further, as their existence would provide the keystone to the next part of the picture puzzle. But they may be still more important, since they may lead to the unraveling of current scientific theory.

Is the current paradigm in danger of being destroyed? I cannot provide a definite answer. In the past, many more challenges to paradigms, or incipient scientific revolutions, have petered out than have succeeded. The odds are against scientific revolutions, and the real conservatism of scientists (which is not always reflected in news media) reflects the paucity of scientific revolutions and the durability of paradigms.

The remainder of this book will explore black holes, quasars, and cosmology, as I examine the possibilities of a scientific revolution and the relations between the model world of the theorist and the real world of the observer. Black holes, furnishing the subject of Part One, are primarily theoretical. Where do they come from? What are they? Have we observed any? Do they really exist? The scene then shifts to quasars and their potential first cousins, the exploding galaxies. Here the observers are ahead of the theorists; they provide observations and explanations sometimes follow. What is a quasar? What type of radiation does it emit? Where does the energy come from? Are the redshifts really cosmological? Quasars, which are probably the most distant objects yet observed, introduce cosmology, the subject of Part Three. Cosmology, the evolution of the entire universe, unifies the rest of astronomy. By studying parts of the universe, we seek to contribute to our understanding of the whole.

This book, like all books about an active research field, is in danger of being outdated by the fast pace of research. I shall identify well-accepted fact, informed opinion, and speculation as such in various parts of the book, and shall summarize these categories in tables at the end of each part. While I try to avoid speculation, I have found that nonastronomers are generally interested in the most speculative parts of the field. Black holes and quasars are strange objects, and you should keep in mind which of their attributes are well understood and which are fascinating products of theorists' imaginations.

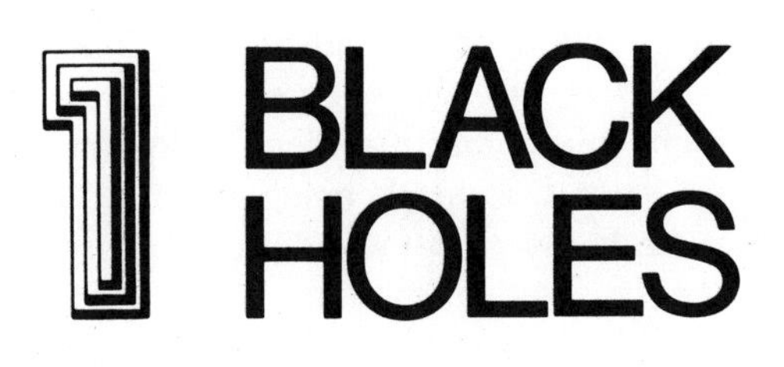

1 BLACK HOLES

Nothing can escape from a black hole, not even light—that is why it is called a black hole. Black holes may form when stars are so overwhelmed by the force of their own gravity that they cannot keep from collapsing. A collapsing star shrinks. If its core is massive enough, it keeps shrinking; the surface keeps collapsing until the entire star has shrunk to a point. Surrounding this point is a volume of space where the gravity is so strong that any light trying to escape to the outside world is sucked back to the central point. No one inside a black hole could communicate with the outside world; he would be cut off from our universe by the event horizon.

Such is our model of the black hole. Is there any reason to believe that black holes are part of the real world, not just exotic inhabitants of the model world? Chapters 2 and 3 concern the known facts of stellar evolution. There are some indications that massive stars do end their lives as black holes, but not enough is known about the late stages of stellar evolution to allow us to say that black holes must exist. Chapter 4 presents a model of a standard, well-understood black hole. There are a few stars, we believe, that may have companions that are black holes, and related bits of evidence that black holes exist in the real world as well as in the model world are discussed in Chapter 5. The speculative frontiers of black hole research are treated in Chapter 6.

STELLAR EVOLUTION: TO THE WHITE DWARF STAGE

Any star spends its entire life in a struggle against gravity. In the same way that we are pulled toward the center of the earth by gravity, every gas atom in a star is pulled toward the star's center. If the star cannot resist this pull by exerting some sort of pressure, the star will collapse as all the atoms respond to this relentless pull and fall inward. If the star kept on collapsing, all the atoms would eventually end up in the center of the star, and the star would become a black hole. A black hole is a dead star, for it will never become anything else. It just sits in space, dark and menacing, swallowing up any matter that comes too close. Fortunately, black holes are very small; there is virtually no chance that the earth will ever collide with one.

How can any stars exist at all, with gravity always trying to turn them into black holes? The interiors of stars must be, and are, able to exert a pressure to oppose the gravitational pull. If the star is to avoid the black hole fate indefinitely, this pressure must sustain itself as the star cools off. Some stars are able to exert such a pressure and end their lives as white dwarfs or neutron stars. White dwarfs compact much of the mass of the star into a volume the size of the earth, while neutron stars are smaller still, only 20 kilometers across.

Do just any stars become black holes? If so, which? How does a star end its life? To further explore these questions, which bear heavily on the origin of black holes, we must consider the factors that govern a star's life cycle. These two factors are gravity and pressure. Different kinds of pressure are at work at different times of the star's life. A star that is fully mature but still robust, the sun, can be our model for the interior of a star.

The battle with gravity

Every atom in the sun is attracted to the sun's center. A balance between heat pressure and this gravitational force enables each atom to resist this attraction. To develop a mental image of this balance, imagine the sun as a ball of gas, 1.39 $\times$ 10^6 km (863,000 miles) in diameter, consisting of three layers: the visible surface, called the photosphere; a gaseous envelope containing most of the mass of the sun; and a small central core where nuclear reactions occur. The entire weight of the envelope presses down on the central core; something must keep the core from collapsing under this tremendous weight.

The weight of the envelope is balanced by the pressure in the central core of the sun. This pressure is tremendous: 2 $\times$ 10^{17} dyn/cm^2, or 2 $\times$ 10^9 times the air pressure at sea level on the earth. It is this pressure that holds the sun up. This pressure comes from the tremendous heat at the sun's center, where the temperature is about 15 million Kelvins.

But can this pressure be maintained? Heat is continuously leaking out of the sun as it shines. The envelope acts as a blanket over the central core as it slows down the heat leakage, but the never-ending flow of energy from the hot core, through the envelope, to the photosphere, and eventually to outer space as the sun shines threatens to deplete the central core of the source of its pressure – heat. How does the sun cope with this heat leakage?

The sun is continuously generating energy in its central core to replenish its heat. This generated energy balances the energy lost into space, so that the sun can continue to hold itself up (Figure 2-1). Energy

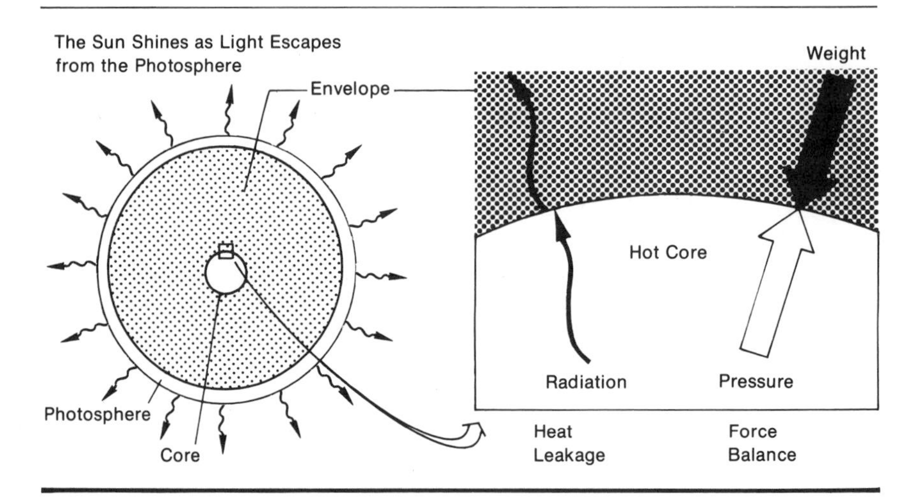

FIGURE 2-1 How the sun keeps from collapsing. Excess pressure in the hot core (white arrow) counterbalances the weight of the envelope (solid arrow). The interior constantly loses energy to the envelope and ultimately to outer space because of the flow of radiation from the core, through the envelope to the photosphere, and to space as the sun shines.

generation maintains the core temperature and the core pressure so that the sun can continue to hold itself up, to shine, and to keep us warm.

The structure of a star like the sun, or the way that it holds itself up, is governed by these two requirements: it must have a sufficiently high central pressure that the weight of the envelope can be supported, and it must have some source of energy in the central core so that the central pressure can be maintained in the face of a continuous loss of energy as the star shines. The energy source is crucial; without it, the weight of the envelope would cause the star to collapse. The life cycle of a star is governed by the evolution of this energy source; the energy source opposes the force of gravity.

One way in which the core can provide energy to keep itself hot is to contract gravitationally. As the core shrinks slowly, with gravity forcing it to occupy a smaller volume of space, gas in it is compressed. As any gas is compressed, it heats up. Heat provided from compression can maintain the central pressure necessary to hold the star up. Yet this energy source is not a long-lasting one.

When a star obtains its energy from gravitational contraction of the core, the star as a whole undergoes changes in its structure because its insides are shrinking. Gravitational contraction cannot supply the energy needed to keep the sun in balance for long, because there is not enough energy to be had from contraction alone. If contraction were the sun's only energy source, the sun would exhaust its energy supply in 15 million years. The earth has been warm enough to support life for a few billion years, and the only source of heat for the earth is the luminosity of the sun. Therefore the sun has maintained the same luminosity for the past several billion years, and for this rate an energy source that lasts only a few tens of millions of years would be insufficient. Clearly there must be some better way to generate energy in the sun's core, some other energy source.

This other energy source is nuclear fusion. If the core is hot enough, with a temperature greater than 4×10^6 K, hydrogen gas can produce energy through fusion reactions, similar to those that make a hydrogen bomb work. In a sense, the sun is one great big slow-burning hydrogen bomb. Following a chain of nuclear reactions, four hydrogen atoms in the sun's core coalesce to form one helium atom and liberate energy in the process. This fusion is a very efficient process and produces a great deal of energy; the conversion of one kilogram (2.2 pounds) of hydrogen to helium yields enough energy to keep a 100-watt light bulb burning for one million years! The heat energy from this reaction in the sun's core keeps the interior hot. As long as hydrogen is being converted to helium in the sun's central core, the sun will remain stable and continue to keep the earth warm.

Evolving stars

While the sun's energy source, hydrogen fusion, produces much energy, it cannot power the sun forever. Other energy sources must come into play, and as these other energy sources become important, the sun's structure will change. Any star will follow the same procedure, though the nature of stellar evolution varies with the star's initial mass, among other things. Eventually any star, our sun also, will exhaust all its energy sources and enter the stellar graveyard, becoming a white dwarf, a neutron star, or perhaps a black hole.

THE MAIN SEQUENCE

Most of the stars in the sky are called main-sequence stars and inside are much like the sun. They keep themselves from collapsing by burning hydrogen in their cores. Different stars, of different masses, may have different detailed structures but their general properties are basically the same. (They differ, for example, in how they transport their energy to the surface and in the details of the fusion reactions.)

The critical factor in distinguishing between various main-sequence stars is mass. Some stars, such as Sirius, the brightest star in the sky, are more massive than the sun. (Figure 2-2 shows where Sirius, along with some other familiar stars, is found in the winter sky.) These massive stars are hotter and brighter than the sun. Sirius, with its mass of 2.2 solar masses, is 21 times as luminous as the sun; if it were where the sun is now, it would be fearsomely bright, and the earth would be too hot to live on. The biggest and brightest main-sequence stars are about 60 to 80 times as massive as the sun and 300,000 times as luminous.

Finding stars less massive than the sun it not so easy, since they are also less luminous than the sun and therefore fainter, and harder to see in the sky. One small main-sequence star, Epsilon Eridani, can be found in the winter sky by tracing out the sinuous outline of Eridanus, the river, southwest of Orion. You need a dark sky and a bit of patience to find it, for Eridanus contains no bright stars visible from midnorthern latitudes. Epsilon lies at a bend in the river, at the western edge of the constellation; its position is shown in Figure 2-2. It is not much further away than Sirius, but since it has only 0.7 solar mass compared with Sirius's 2.2, it is much fainter, with a luminosity that is 0.25 the sun's. It is so faint that it does not deserve a proper name; it is called Epsilon since

FIGURE 2-2 The southern winter sky as it appears from midnorthern latitudes. The brighter stars are shown, along with guideposts to the fainter stars in Eridanus.

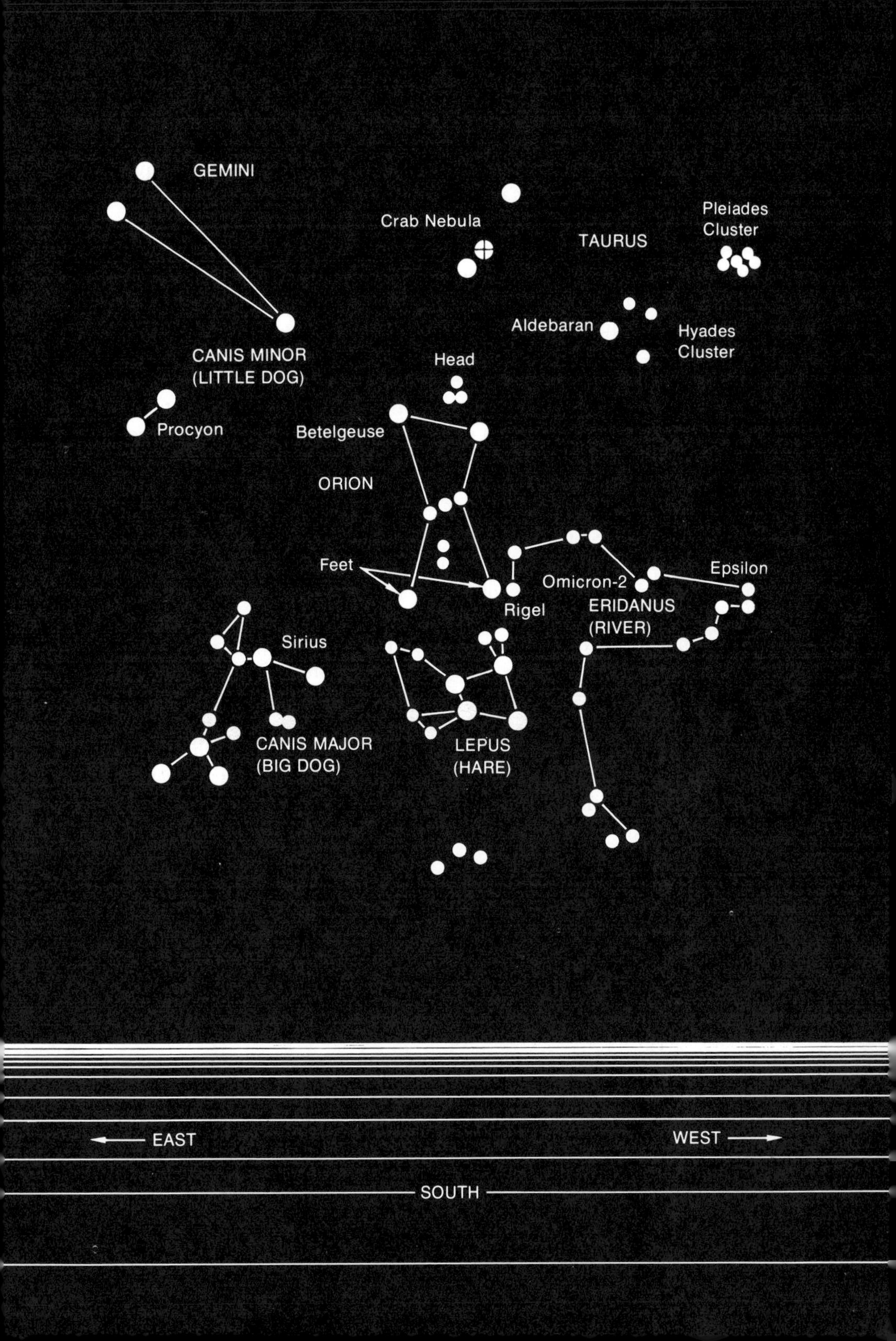
GEMINI
Crab Nebula
TAURUS
Pleiades
Cluster
Aldebaran
Hyades
Cluster
CANIS MINOR
(LITTLE DOG)
Head
Procyon
Betelgeuse
ORION
Feet
Omicron-2
Epsilon
Rigel
ERIDANUS
(RIVER)
Sirius
CANIS MAJOR
(BIG DOG)
LEPUS
(HARE)
EAST
WEST
SOUTH

it is the fifth brightest star in the constellation of Eridanus. (*Epsilon* is the fifth letter in the Greek alphabet. The brightness ordering was done by eye estimates and is not always precise.)

In the winter sky we can see bright, massive, hot main-sequence stars like Sirius and faint, small, and cool main-sequence stars like Epsilon Eridani. Why? The luminosity of a main-sequence star depends almost entirely on its mass. In a massive star, such as Sirius, the interior must support the weight of the envelope by maintaining a very high pressure. The interior must be quite hot so that the pressure will be kept high enough. As a result the interior contains much heat energy that leaks out into space and makes the star shine brightly. The nuclear furnace must burn rapidly to replenish this leaking energy, and the hot core produces fast-burning nuclear reactions. In a less massive star, like Epsilon Eridani, the interior is cooler than Sirius's, since there is not so much weight to support. The pressure is less, the temperature is lower, the nuclear reactions proceed more slowly, there is less energy to leak out into space, and the star is dimmer.

This picture of a galaxy of main-sequence stars – some massive, hot, and luminous; some small, cool, and less luminous – does not depict the whole story of stellar evolution. Stars stay on the main sequence only as long as they can burn hydrogen at their centers. Sooner or later, as a star burns all its hydrogen to helium ash in its central regions, its core will run out of hydrogen fuel. Massive stars, in which the central nuclear reactions proceed rapidly, will run out of fuel long before their less massive counterparts. The sun will remain on the main sequence for 5 billion years; and Sirius, a more massive star, will leave the main sequence 1.5 billion years after its birth. All massive stars are young; massive stars burn themselves out to the point of invisibility before they can become old. But what happens to these stars that leave the main sequence when they run out of hydrogen at their centers?

RED GIANTS

Before stars die, they go through some very interesting evolutionary stages. A star that has run out of hydrogen fuel at its center leaves the main sequence and turns to the only other available source of central energy, gravitational contraction of the core. As the helium core contracts, it heats up, thus producing energy to replenish the energy lost as the star shines. Paradoxically, as the core and interior of the star shrink and heat up, the envelope expands and cools. (The details of this process are complex and fortunately peripheral to the story.) The star's surface cools to less than half of its main-sequence surface temperature, and the star swells to tens or hundreds of times its main-sequence size, while the interior is shrinking and increasing in temperature. The star becomes a red giant.

A red giant is continuously readjusting itself to maintain the necessary balances as the core shrinks. Eventually the core becomes hot enough that another nuclear reaction can begin at the center: the helium nuclei, the ashes left over from the earlier main-sequence hydrogen burning, now become fuel as they fuse to become carbon nuclei. (In a red giant, hydrogen is still being fused to helium in a shell, but the shell cannot provide energy to hold the inner parts of the star up.) The core must be hotter than 10^8 K to make the helium-to-carbon reaction go. But once the star starts fusing helium, it has found another way to keep its outer envelope from collapsing, and the contraction of the core stops, for a while at least.

One of the best-known red giant stars is Betelgeuse, the bright star in Orion's right shoulder. Orion is an easily recognized constellation, a bastion of the winter sky. (Figure 2-2 shows where it is. Yes, the name Betelgeuse is pronounced "beetle juice." It comes from the Arabic *Ibt Al-Jauzah,* meaning "Armpit of the Central One.") The surface of Betelgeuse is cool, only "red" hot (about 2300 K), whereas the sun is "yellow" hot (5760 K). Betelgeuse varies in size, expanding and contracting irregularly over periods of a year or so. It is truly a giant star. If you should put the center of Betelgeuse where the sun is, the earth would always be inside Betelgeuse's surface and would be vaporized (Figure 2-3). The sun, too, will become a red giant in 5 billion years, when it exhausts its hydrogen fuel. It will not be as large as Betelgeuse,

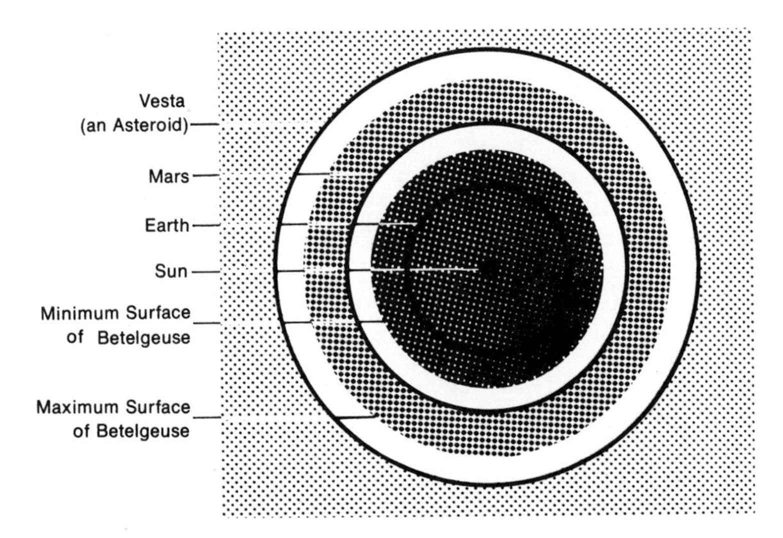

FIGURE 2-3 The size of Betelgeuse, compared with the solar system.

but the earth will be uninhabitable nevertheless, since the sun will be too bright. We have 5 billion years to prepare for this fate, so the human race should be able to find a new place to go by then. Perhaps one of Jupiter's moons could be made quite pleasant. . . .

The center of Betelgeuse is much hotter than the center of the sun. The helium fuel in the former has probably been exhausted, leaving a core of carbon. It is probably burning helium in a shell surrounding the hot, shrinking carbon core, with a central temperature of several hundreds of millions of degrees, in contrast to the sun's comparatively placid 15 million degrees. The sun, a middle-aged star, still has fuel left for 5 billion more years on the main sequence. Betelgeuse is old, doing its best to survive on the remaining fuel. It is rapidly approaching the day of reckoning when it will totally exhaust its nuclear energy supply.

What happens to stars like Betelgeuse as they near the end of their stellar lives? During the red giant stage, the star's core is contracting under the influence of gravity. Every once in a while, new nuclear reactions can begin as the center becomes hot enough. Helium burns to form carbon and oxygen; carbon fuses into neon and magnesium; oxygen burns to silicon and sulfur; and neon, magnesium, sulfur and the rest fuse in a series of partially understood reactions to form iron. Once you make iron, no more energy can come from fusion. If you wish to fuse iron with any other atom, you must supply energy; you cannot use such reactions to supply the energy needed to hold a star up. This contraction is shown in Figure 2-4; the star's evolution is indicated toward the right of the diagram, showing increasing compaction of the center, or higher central density, caused by gravity. The star is shown stopping every once in a while on one of the lines as a reaction takes place that temporarily delays the collapse.

The details of stellar development during the late red giant stage are still poorly understood. The overall picture is clear. Gravity sends orders to the core: the core must contract. Execution of these orders can be temporarily postponed when nuclear reactions can supply energy to the core and keep the interior hot enough for the weight of the envelope to be supported. However, any fuel that the star turns to as an energy source is eventually exhausted. The core again contracts, and the star's evolution proceeds inexorably to the final state.

REAL STELLAR EVOLUTION

A model is thus known for the life of a star from main sequence to red giant. How much confidence do we have that real stars actually evolve this way? Does this model of stellar evolution, which sounds so neat, have anything to do with reality? We must bring the observations

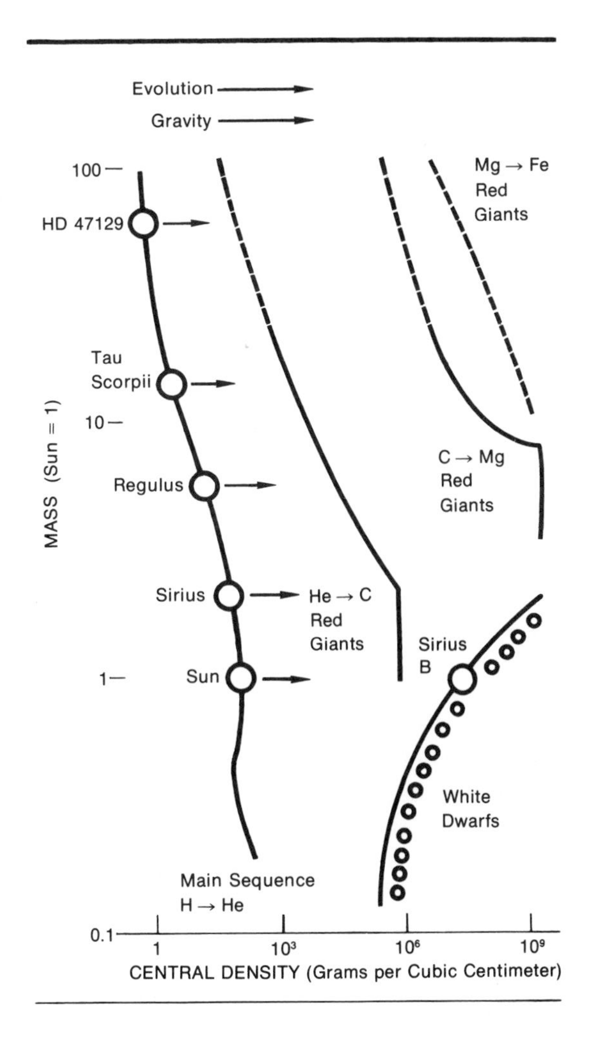

FIGURE 2–4 Stellar evolution from main sequence to white dwarf. Gravity always compresses the star, carrying it toward the right of the diagram as the central density increases. If the star does not lose mass, it travels in a horizontal line to the right-hand side of the diagram. If a star loses mass, it moves downward. The positions of these curves are very uncertain for stars heavier than 15 solar masses. Dashed curves are extrapolated from published theoretical models.

in. The model has been checked extensively and generally holds up well; representative examples are convincing.

The model must provide that the mass of a star determines the star's evolution, at least through the red giant stage. In particular, a massive star will spend less time on the main sequence than a less massive one. To check this provision, consider two clusters of stars in the constellation Taurus, seen in the western sky in winter. These are the Pleiades, or Seven Sisters, and the Hyades, which surround the bright star Aldebaran. (Aldebaran is not a member of the Hyades cluster; it just happens to lie in the same direction in space. See Figure 2-2 for the location of these objects.) The Pleiades are young stars, and the cluster contains many bright, blue, high-mass main-sequence stars. Such stars do not exist in the older Hyades; they have burned out. The observations fit the model.

Another expectation of the model is that massive stars should be more luminous than less massive stars. Earlier, the sun and Sirius were compared; Figure 2-5 shows how theory and observation compare for a larger number of stars. The agreement seems good, except for stars in the Hyades. The Hyades may differ from the other stars in chemical composition, which would cause their luminosities to be different. Or, it has been proposed, our estimate of the distance to the Hyades may be wrong, and the masses and luminosities of these stars as we accept them now may not be correct. (This alternative is gaining ground.) But generally, agreement between theory and observation is good.

There is a potential time bomb ticking away on the shelves of the stellar evolutionists. Theory indicates that there should be a detectable number of neutrinos coming from the sun. These neutrinos are subatomic particles produced in nuclear reactions at the sun's center, and they interact with matter so rarely that they pass right through the sun from center to surface. (A neutrino beam would have to pass through several hundred light-years of lead to be completely stopped.) Seeing these neutrinos would give us clear evidence about the sun's center.

Dr. Raymond Davis has attempted to find these neutrinos by intercepting them with a swimming-pool-sized vat of cleaning fluid located deep in a South Dakota gold mine. So far, he has had no success. His results indicate that the number of neutrinos from the sun is less than one-fifth of what the theory provides for. Whether there is some minor deficiency in the theory or whether there is a significant weakness in the model is not known at present.

The observational tests of the theory described above all relate to main-sequence stars. It is not so easy to check the theory against red giant stars, since there are fewer red giants in the sky. Furthermore, most of the interesting changes in a red giant take place in the giant's invisible core, and their effect on the surface of the star is not straight-

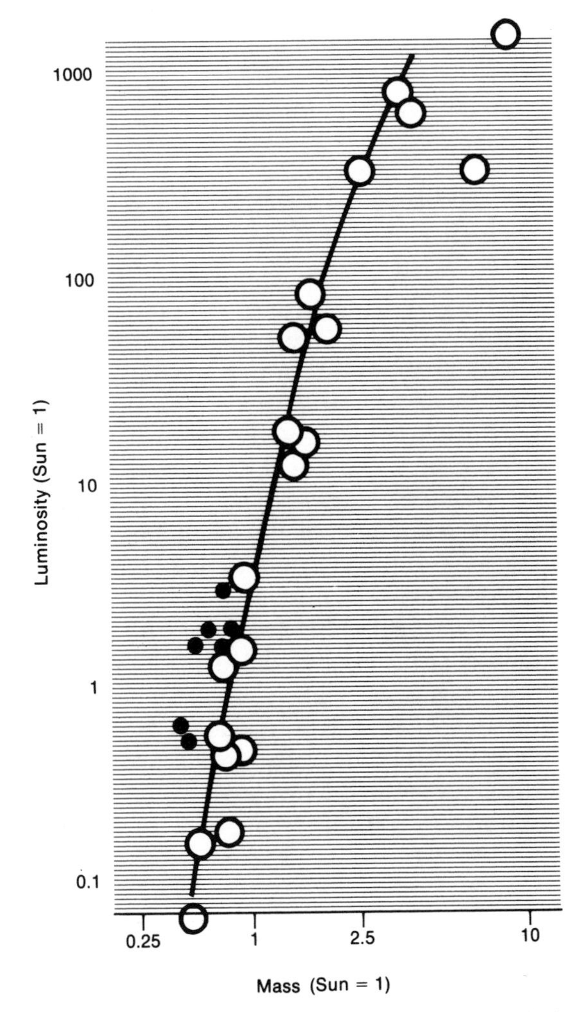

FIGURE 2-5 The mass-luminosity relation. The solid line is the theoretical relation for main-sequence stars of 60% hydrogen, 27% helium, 3% heavier elements. The circles mark nearby main-sequence stars; the dots mark the Hyades. (Adapted from I. Iben, Jr., *Astrophysical Journal,* vol. 138, 452, 1963, published by the University of Chicago Press. Copyright © by the University of Chicago. All rights reserved.)

forward. However, red giants in clusters, as far as they can be checked, have the luminosities and temperatures that theory says they should have.

The fate of stars beyond the red giant stage is still somewhat mysterious. Eventually, a star will finish its nuclear burning and become a corpse, an object that is generating no energy. Scientists have observed two types of stellar corpses: white dwarfs and neutron stars. A third may exist, black holes. The white dwarfs, you will recall, are objects in which the mass of the star has been packed into a planet-sized volume.

Stellar corpses: white dwarfs

We left the red giant as its core was continually following the orders of gravity to contract, stopping occasionally to ignite and complete a nuclear reaction in its core. Must the orders of gravity be followed? If so, the star must continue to contract indefinitely, becoming a black hole. As the star reaches the end of its life, heat pressure eventually loses the battle with gravity — the energy needed to sustain the heat has been lost into space as the star shines. When the fuel inevitably runs out, the star's interior cools to a point at which heat pressure is no longer important. But if a small enough core is left after the red giant stage, a star can find a final resting place as a white dwarf, a star no larger than the earth (Figure 2-6).

The first white dwarf ever discovered, and the brightest known today, is Sirius B, also called the Pup, Sirius being the Dog Star. Although Sirius looks like a single star to the naked eye, a good telescope will show that it is a system of two stars orbiting each other. The brighter of the two, called Sirius A or often merely Sirius, is the main-sequence star mentioned earlier. It is difficult to see the companion, known as B, since its feeble light is overwhelmed by A; Sirius A is ten

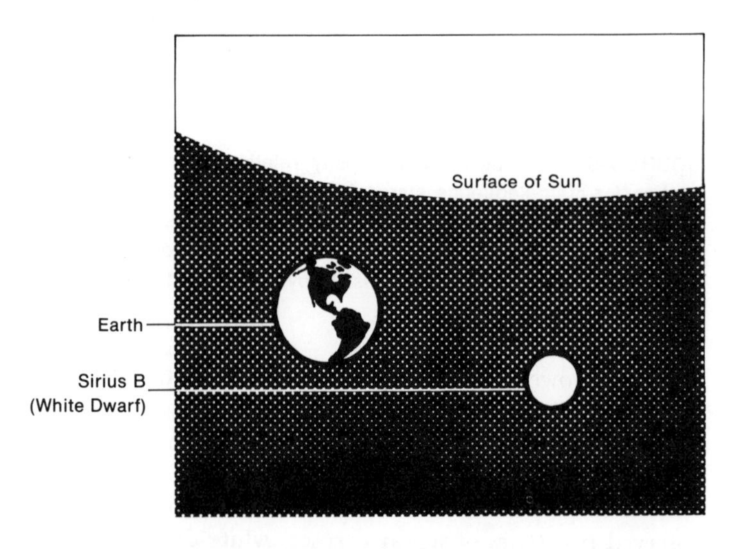

FIGURE 2-6 White dwarfs are very small.

thousand times brighter than B. Sirius B is faint because it is small, packing into an earth-sized volume the same mass that the sun contains in a much bigger volume. Thus Sirius B is much denser than the sun. A matchbox full of solar matter would weigh about 15 grams (half an ounce), while the same volume of Sirius B matter would weigh about 10,000 kilograms (10 tons) if it were weighed here on earth (Figure 2-7).

HOLDING WHITE DWARFS UP

Sirius B, the Pup, has already passed through the red giant stage. How has it managed to defeat the relentless command of gravity? Density is the key. Sirius B manages to keep itself from shrinking because its interior exerts a kind of pressure that has no connection with heat—degeneracy pressure. The name comes from the degenerate condition of the electrons in a high-density state. (It has nothing to do with morals, as electrons do not have morals.) Degeneracy pressure does not originate from heat; it originates from density alone. As a result, Sirius B doesn't mind cooling off. It can still hold itself up because the necessary internal pressure does not depend on heat. Even if Sirius B were stone cold, it would not collapse.

The magic of degeneracy pressure comes from the combined factors very high density of white dwarfs and internal properties of electrons. In an ordinary gas, such as the gas in the center of the sun (Figure 2-8 left), atoms are so far apart that their size is much smaller than the

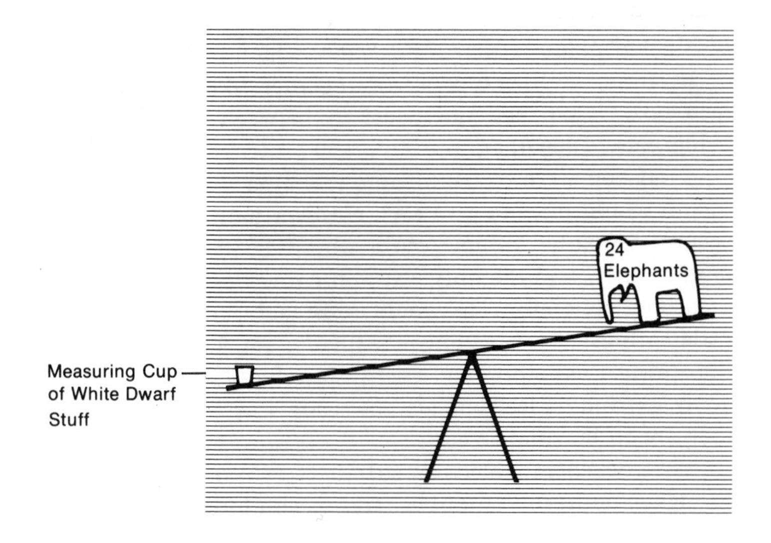

FIGURE 2-7 White dwarf stuff is very dense; a cupful would outweigh two dozen elephants.

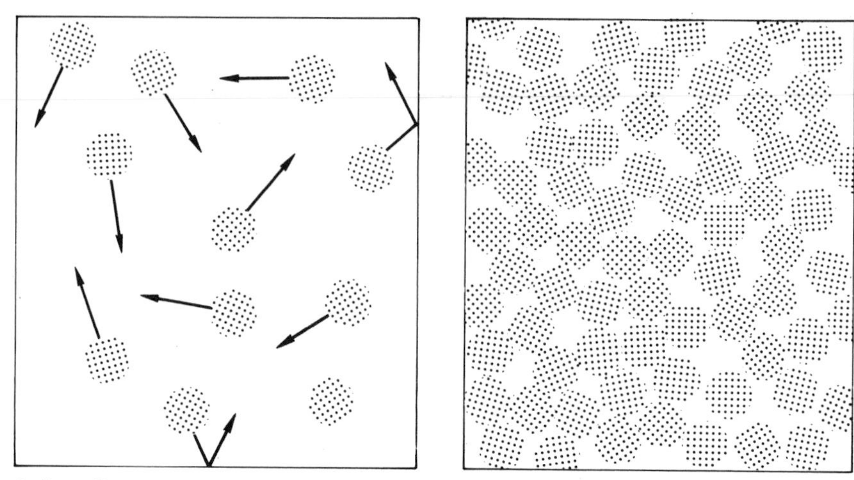

FIGURE 2-8 In an ordinary gas (left), atoms are free to move about, since they are much smaller than the distances separating them. When the gas is so dense that the electrons become degenerate, the packing of the electrons exerts pressure (right).

distance between them. They can be visualized as point-particles. They fly about freely and exert pressure by bouncing off other particles. In a degenerate gas, atoms are packed so closely that they can no longer be considered point-particles. Electrons, which make up most of the volume of an atom, are being squeezed together like little balls packed tightly in a box. There are no spaces in between the electrons. They resist being packed together.

If you were to try to compress a gas with degenerate electrons in it, you would have to give each electron more energy as you forced it to occupy a smaller volume. You would be trying to pack the electrons more tightly, and the only way you could do this would be by squeezing each one. The electrons would resist being squeezed, in the manner of little springs. As you squeeze a spring, you have to supply energy to it; the spring then stores the energy internally. If you let go, it will suddenly expand. Electrons act the same way, demanding more energy if they are to be compacted.

A degenerate electron gas acts very much like a collection of little springs. You have to supply energy as you force the electrons (or springs) to occupy a smaller volume, and this energy comes from the force you exert to compress the gas. *Degeneracy pressure* is the term for the pressure the gas exerts, opposing the force that compresses it. It is this resistance to compaction, which comes from the need to supply compacted electrons with more energy, that holds a white dwarf up.

The critical aspect of degeneracy pressure, from a stellar evolutionist's point of view, is that it has absolutely nothing to do with temperature. Temperature describes the motion of the atoms of a gas, and heat pressure comes from the motion of these atoms. For heat pressure, the higher the temperature, the faster the movement of the atoms. They hit each other harder, and the pressure is higher. In a degenerate gas, the atoms are packed so closely that their collisions are unimportant; they cannot move very much. Thus, from the point of view of pressure, whether the gas is hot or cold does not matter very much. The gas can cool off to any degree but the pressure will still be there.

WHITE DWARF STRUCTURE

Referring once more to stars, let us see how degeneracy pressure keeps a white dwarf from collapsing. A white dwarf is a sphere of degenerate matter. Electron degeneracy holds the star up, as it enables white dwarf matter to resist being compacted by gravity. The degeneracy pressure is completely independent of temperature. Thus as the white dwarf cools, it remains the same size. Ultimately it will become a black dwarf, but current ideas about the cooling rates of white dwarfs indicate that the universe has not existed long enough to have allowed any of them to cool to less than a few thousand degrees.

Degeneracy pressure acts like a barrier, preventing a white dwarf from further collapse (see Figure 2-4). Once a star has become a white dwarf, it can collapse no further. Not all stars become white dwarfs, as degeneracy pressure has its limitations. If a star with more than 1.4 solar masses attempts to stabilize itself as a white dwarf, electron degeneracy will be insufficiently strong to hold it up and the star will collapse further. The existence of this limit, known as the Chandresekhar limit, means that stars that are to become white dwarfs must end their lives with less than 1.4 solar masses of material. Although rotation could, in theory, stabilize a heavier white dwarf, most known white dwarfs do not rotate fast enough to allow rotation to be significant.

The white dwarf model can be checked. If you know the mass of a white dwarf, the theory tells you its density (it can be read off a graph like Figure 2-4). Knowing its density tells you how big the white dwarf should be: high-density white dwarfs are small; low-density ones are large. You can then go and measure sizes and masses of white dwarfs and see how well they conform to the theory. These measurements have been made and vindicate the theory.

Origin of white dwarfs

We have lately examined a stellar corpse — a white dwarf, and a star just about to die — a red giant. How are these two stages of stellar evolution connected? Unfortunately, the late evolution of red giants is a complex phenomenon and all the factors are not understood. But there is a theory, which the Polish astronomer B. Paczynski and the American William Rose developed, for the origin of white dwarfs. I present this theory in the remainder of this chapter and then discuss the evidence for it. To this point, this chapter has been factual: well-tested theories and observations. From here to the end, it is informed opinion: ideas that a number of astronomers (including me) believe to be true but that have not been sufficiently well proved to be accepted unequivocally.

Let us consider the past life of Sirius B, the Pup, in the Paczynski-Rose scenario for the evolution of white dwarfs. At one time the Pup was a red giant, considerably larger and brighter than Sirius A. The Pup would then have been brighter than Venus at its brightest — bright enough to cast a shadow here on Earth. It then had more mass than Sirius A now has, probably 2.5 to 3 solar masses, contrasted with A's 2.2 solar masses. The Pup's core then contracted as it successively used up various nuclear fuels. Eventually the core became so dense that degeneracy pressure from electrons could prevent the further collapse of the star. At this point, the Pup still looked like a red giant, but was basically a very hot white dwarf core, with a mass of one solar mass, surrounded by a gigantic, very tenuous envelope, which contained the remaining mass.

Eventually, in Paczynski's picture, the core lost its grip on the envelope, as pressure from radiation from the core overbalanced the pull of gravity. The envelope drifted off into space, becoming visible as a gas cloud surrounding a small, hot white star. We can see such gas clouds, known as planetary nebulae, around other stars; one of these is shown in Figure 2-9. Rose believes that planetary nebulae are formed somewhat more forcefully, as explosions in the hydrogen-burning shell of the star cause the envelope to blow away. In either case, the red giant loses its outer envelope, violently or nonviolently, leaving behind the core, which contains roughly half the mass that the star started with.

The envelope then dissipated in a few thousand years, a very short time as far as stellar lives are concerned, and the white dwarf core, the star we now know as the Pup, was left. The Pup cooled as its atoms and electrons gradually lost their energy of motion. However, since the Pup's core was supported not by the motion of the electrons (heat pressure) but by electron degeneracy pressure, the Pup did not contract any more and remained constant in size. It is now quite hot, some 33,000 K, three

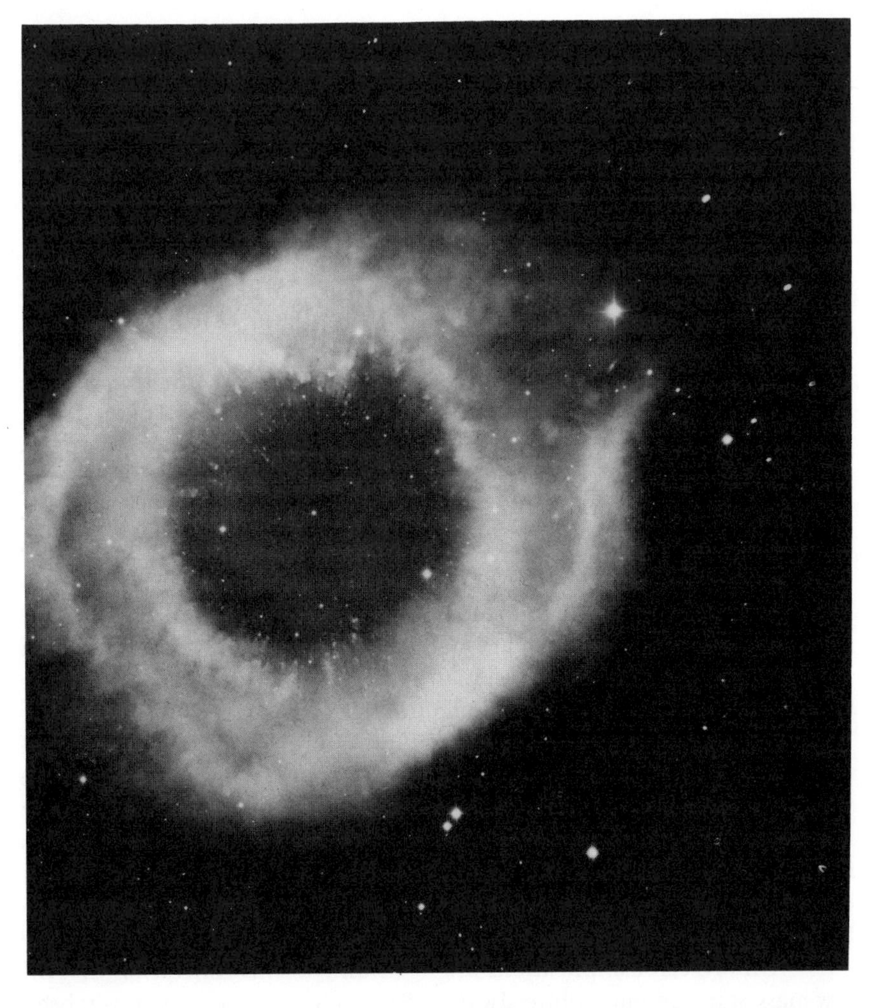

FIGURE 2-9 NGC 7293, a planetary nebula in Aquarius. (Hale Observatories.)

times as hot as Sirius A, but very small – smaller than the earth. In the future, it will continue to cool, becoming a black dwarf some tens of billions of years from now. It will be the same size as it is now but will emit no light.

The sky is full of white dwarfs; tens of thousands of white dwarf suspects exist and several hundred have been confirmed as white dwarfs, their spectra having been obtained. By coincidence, Figure 2-2 shows three of the four white dwarfs nearest to the earth. Procyon, in the constellation Canis Minor (Little Dog), is a double star somewhat like Sirius in having a white dwarf companion. Omicron-2 Eridani, another star in this dim constellation, is a triple system, with one white dwarf member. All white dwarfs have masses of less than 1.4 solar masses; they all have roughly the same diameter; and they are all quite faint. None of the three white dwarfs just mentioned could be seen as a single star with the naked eye; their brighter, main-sequence companions allow the stargazer to see where they are.

Do all stars become white dwarfs? If so, black holes would exist in the model world only, as gravity could not make any star smaller than a planet. Black holes would then pass into the world of esoteric mathematics and out of our immediate consideration. But it is only the small stars that become white dwarfs.

The reason that all stars may not become white dwarfs is that some stars are too massive. Remember, degeneracy pressure can be asked to hold up only a star smaller than 1.4 solar masses; eventually in larger stars the electrons are squeezed to the point that degeneracy pressure can no longer cope with gravity. Gravity can force a star around the white dwarf barrier to collapse (shown in Figure 2-4) if the star finishes its life with a mass of more than 1.4 suns. Sirius B could never have become a white dwarf if it had not unloaded 1.5 to 2 solar masses of gas when it shed its envelope as a planetary nebula. Thus the only stars that can become white dwarfs are those small enough to end the red giant stage with cores smaller than the magic figure "1.4 solar masses."

But saying that a white dwarf forms from a core of less than 1.4 solar masses begs the question. We seek to connect the red giant stage with the white dwarf stage a little more securely and really to answer the question, What stars become white dwarfs? Paczynski's calculations indicate that small stars, stars with less mass than some critical mass, become white dwarfs. Latest results indicate that this mass limit is between 2 and 6 solar masses, with a best guess of 3.5.

Unforeseen complications might invalidate this conclusion that all low-mass stars become white dwarfs. Several factors have not yet been included in calculations like Paczynski's. Magnetic fields and rotation may be important. Stars in close binary systems probably evolve quite differently. But it does seem probable that at least some stars follow the hypothetical sequence outlined above for Sirius B: red giant ⟶ planetary nebula ⟶ white dwarf.

Observational support for this scheme comes from two facts. The number of white dwarfs in our galaxy is roughly equal to the number of stars smaller than 3.5 solar masses that have evolved past the red giant stage. Stars at the center of planetary nebulae resemble hot white dwarfs, and theoretical calculations indicate that they should eventually become white dwarfs. The sequence low-mass star ⟶ red giant ⟶ planetary nebula ⟶ white dwarf is one possible end to the life of a star.

Stars are gigantic spheres of gas. A star's structure is determined by a star's need to hold itself up by some sort of internal pressure. Evolution of a star is governed by the changing nature of the processes that maintain this pressure. A main-sequence star obtains its energy from the conversion of hydrogen to helium in its center. When the star runs out of central hydrogen, it becomes a red giant as its core contracts and its surface expands. The red giant turns to other nuclear reactions in the center as it searches for a way to hold itself up, but it eventually runs out of all possible fuels. If the core is small enough, the core can stabilize itself as a white dwarf, since degeneracy pressure from electrons can hold the star up without involving heat. This white dwarf stage is one possible end to stellar evolution: a star with roughly the mass of the sun squeezed into a volume no larger than the earth's. Do all stars end their lives as white dwarfs? No, for we have observed another endpoint to stellar evolution: the neutron star or pulsar, in which one solar mass or so of gas is compressed into a tiny sphere 20 kilometers across.

SUPERNOVAE, NEUTRON STARS, AND PULSARS

> In the first year of the Shih-huo period, in the fifth moon, on the day of Ch'ih Ch'iu [July 4, 1054], a guest star appeared several inches southeast of T'ieng Kuang [a star in what is now called the constellation of Taurus].
>
> — *Sung-Shih* (History of the Sung Dynasty)

> I make my kowtow. I observed the phenomenon of a guest star. Its color was slightly iridescent. Following an order of the Emperor, I respectfully make the prediction that the guest star does not disturb Aldebaran [the brightest star in Taurus, Figure 2-2]; this indicates that . . . the country will gain great power. I beg to store this prediction in the Department of Historiography.
>
> — Yang Wei-T'e, Imperial Astronomer, 1054

So did the Chinese court record the appearance of the most spectacular event of stellar evolution, a supernova. They saw a star appear where no star had been seen before. What was this — the birth of a new star? No, there *had* been a star there; it was merely too faint to be seen with the naked eye. Supernovae mark the death, not the birth, of stars. Suddenly this dying star became much brighter — as bright as the whole Milky Way galaxy. It could be seen in daylight for several months, and in the nighttime for a year or so. "Eventually it faded, and became invisible."[1] (Notes are found at the end of the book.)

The Chinese astronomers started a story that culminated some nine hundred years later in the discovery of neutron stars — objects whose extreme density can be understood by imagining the entire mass of the sun could be packed into a volume the size of the earth's crust under a typical U.S. county, only 20 kilometers across. We have identified the Crab Nebula (Figure 3-1; Figure 2-2) as the debris of this explosion that the Chinese recorded in 1054. This cloud of glowing gas is about three parsecs across and is filled with electrons gyrating around magnetic lines of force at speeds close to the speed of light. Near the center of the nebula is this neutron star, which was recognized as such only in 1968. Neutron stars make white dwarfs look almost normal, as they are 10^9 — one billion — times as dense as this other form of dead star.

FIGURE 3-1 The Crab Nebula, Messier 1. This is the remnant of the 1054 supernova. (Hale Observatories photograph.)

Neutron stars are the second exhibit in the gallery of stellar corpses. The existence of such objects was first proposed in the 1930s, shortly before it was realized that black holes also could exist. Neutron stars and black holes both remained in the speculative fringes of the model world until the 1960s, when the discovery of pulsars brought the neutron stars into the real world of discovered astronomical objects. They are part of the violent universe, since they are the debris of a supernova explosion. In order to ask whether there is still a third type of stellar corpse — the black hole — we must first find out what neutron stars are and where they come from.

The tortuous tale of how these objects were discovered illustrates the interplay between the model world and the real world, and the story may be repeated with the discovery of black holes. The neutron stars themselves may tell us something about the types of stars that created them. When we have established the neutron-star story, we can extend the evolutionary scenario into a grand scheme for the death of stars, the current working hypothesis for the end of stellar evolution. While this scheme may not be correct, it is a useful way of organizing what we do and do not know about the way that stars die.

Exploding stars

The Chinese chronicles quoted at the beginning of this chapter referred to a "guest star," or one that appeared where no star had been seen before. The appearance of a guest star is quite a surprise to anyone who is accustomed to looking at the sky. The familiar stars in their familiar patterns — the constellations — have seemed to be immutable. Year after year, the same stars rise at their accustomed time. You always see the Pleiades, part of Taurus (Figure 2-2), rise around midnight in July, and they are followed by the bright star Aldebaran. Thus it always was and thus it will always be, or so it seems. Occasionally though, the familiar patterns of the stars are disrupted by the appearance of a new star — a bright one, out of place. One such new star was the Chinese "guest star" of 1054, in Taurus. Tycho Brahe, one of the key figures of the Copernican revolution, made his reputation by carefully observing one of these celestial interlopers, the supernova of 1572. His book *De Nova Stella* ("On The New Star") gave Tycho his reputation and these new stars a name: novae, or new stars. Two classes of "new stars" are now recognized, the *novae* and the much brighter *supernovae*.

NOVAE AND SUPERNOVAE

The names *novae* and *supernovae* are misleading, since these stars are not "new" stars at all but stars that brighten spectacularly as they leap to the stellar graveyard. What is the difference between these two classes of new stars — novae and supernovae? The answers to these questions deepen our probe into the late stages of stellar evolution.

The ordinary novae are much more common but much fainter than the supernovae, which are more spectacular. Novae that are visible to the naked eye occur every decade or so, and they rarely become as bright as the brightest star in the sky. If you look for a nova, you just look for an extra star in some constellation. The luminosities of novae are comparable to the luminosities of the brightest stars in the galaxy — up to 10^6 times the luminosity of the sun.

Supernovae are much rarer; the last one seen by men on earth was Kepler's supernova of 1604. They are much more powerful than the novae; they become as bright as an entire galaxy – billions of times as luminous as the sun. Historic supernovae in our galaxy have been visible in the daytime for as long as two months.

The ordinary novae are related to the white dwarf sequence of stellar evolution. Some of them are recurrent; they have been known to flare up more than once. The nova phenomenon generally occurs in a binary system that has as one member a very hot white dwarf. A detailed theoretical model of novae that agrees with all observations does not exist yet, but it is generally supposed that gas falling on the surface of the white dwarf triggers the nova outburst. This gas is pulled from the other star by gravity. While there are many unanswered questions about novae, it seems probable that the nova phenomenon is one branch of the white dwarf sequence. The supernovae are similar to the novae, in that the same general pattern of a star's becoming vastly brighter occurs. Yet they are different: once a star has become a supernova, it has died.

SUPERNOVAE IN HISTORY

Historic observations of supernovae are few, because the phenomenon occurs infrequently. Most of the supernovae in our galaxy were recorded by the Chinese, whose court astronomers, careful observers of the sky, recorded the appearance of unusual objects. Their thoroughness is attested to by the fact that they did not miss any appearances of Halley's comet in the last two thousand years. You go back through the chronicles and faithfully every 76 years the comet's arrival is recorded. They did distinguish between comets and "guest stars," or novae and supernovae. Somewhere around half a dozen supernovae have been recorded in the chronicles. We have gained much information about the brightness of supernovae from the chronicles, but sometimes it is a little difficult to figure out exactly what the chronicles mean. For instance, the A.D. 185 supernova is described: "Kheihai of the second year of Chung-P'ing [early Han dynasty], a guest star appeared in Hang Mang, about the size of half a mat. It was of five colors. . . ."[2] What is the visual magnitude of half a mat? We seek quantitative information; the existence of a supernova is a useful datum, but it would be nice if we could know precisely how bright the A.D. 185 supernova was.

The two supernovae seen in Europe, during the Renaissance, were more precisely recorded, since they were observed by Tycho Brahe, the founder of Western observational astronomy, and his successor, Johannes Kepler. Characteristically, Tycho noted the supernova's brightness as contrasted with other stars in the constellation Cassiopeia; we can thus reconstruct quantitatively how bright the supernova was. Kepler made some observations of the 1604 supernova and collected the records of others, so the 1604 supernova is usually called Kepler's supernova. Since then, there has been no observation of a supernova in our galaxy.

Serious work on supernovae did not begin until the 1930s, when Fritz Zwicky, the pioneer of supernova research, realized that he must look toward other galaxies to see supernovae with any substantial frequency. In 1933, he began examining nearby galaxies with a 10-inch refractor to see whether any stars had flared up. In 1936, his chances for finding supernovae improved when the first Mount Palomar telescope, an 18-inch Schmidt, was installed to begin the Palomar supernova search program. The telescope still patrols the sky, examining galaxies to see if any stars in the galaxies suddenly become brighter. In 1939, the search program found the brightest supernova of the twentieth century. This supernova, shown in Figure 3-2, was a ninth magnitude star at maximum. The brightness of this supernova allowed CalTech astronomer Rudolf Minkowski to obtain a magnificent series of spectra of this supernova; through these and similar series of spectra, we have some understanding of the supernova phenomenon.

SUPERNOVAE AND STELLAR EVOLUTION

What causes a star to become a supernova? Why should a star suddenly become as bright as an entire galaxy? A supernova explosion is a stellar funeral, marking star death. The star's inner core collapses, for the star has burned all its nuclear fuel and is unable to hold itself up. The collapse of the core releases vast amounts of energy, which cause the

FIGURE 3-2 The supernova in IC 4182, the brightest supernova of the twentieth century. (Hale Observatories photograph.)

envelope to expand. Most of the details of this process are still mysterious, but the general outlines are understood empirically even if we cannot model them.

We are somewhat more confident of our picture of what happens to the outside of the star after these events in the interior have triggered the supernova explosion. The violence in the interior causes the outer surface of a star to be shot off into space with a velocity of several thousand kilometers per second. This gas, the outer part of the star, is heated by the explosion. As it expands, it becomes less dense, so that a shell of hot gas expands out from a central core. The spectrum of this gas shell qualitatively resembles the spectra of other clouds of hot gas, seen in our galaxy. Quantitative analyses of these spectra are still rudimentary, as the instruments needed to observe the spectra of faint objects like supernovae in extremely distant galaxies have been developed only recently. We still do not know the answers to some critical questions. What is the composition of this gas? How hot is it? What fraction of the star is shot out into space, and what fraction remains in the central core?

As the supernova fades, a month or so after the explosion, the gas shell becomes larger and more rarefied. After some thousands of years, this gas cloud is several parsecs across. We see several of these clouds in the galaxy, including the Crab Nebula.

Clearly our knowledge of what happens during a supernova explosion is sketchy. The power of modern instrumentation is no match for faintness, and recently seen supernovae are faint because they occur in distant galaxies. If one were to occur in the Milky Way, or even in a relatively nearby galaxy, we should be able to learn much more. We are waiting. But now our attention focuses on the object left in the middle of that expanding cloud of gas. This object is the neutron star, a second way that Nature has managed to defeat the orders of gravity for the star to collapse indefinitely.

Pulsars: the neutron star discovered

Supernova pioneer Fritz Zwicky and his colleague Walter Baade put forth a far-reaching but speculative suggestion in 1934: "With all reserve we suggest the view that supernovae represent the transitions from ordinary stars into *neutron stars,* which in their final stages consist of extremely closely packed neutrons."[3] Five years later, in 1939, J. Robert Oppenheimer (who later became well known in connection with the atomic bomb) and his student George M. Volkoff showed that yes, indeed, neutron stars *could* exist. These objects are about 20 kilometers across, the size of a large city (Figure 3-3). At their centers they are 10^{15} times as dense as water — as dense as the nucleus of an atom.

But all this work was purely theoretical. It is one thing to suggest the possible existence of something that packs the mass of the sun into a volume 20 kilometers across, and another thing actually to find such an object. The work of the 1930s only showed that neutron stars could be inhabitants of the model world; before they could enter the real world, they had to be found.

Not much happened in the neutron star field until the 1960s. Supernova work was continuing, of course. During World War II, when

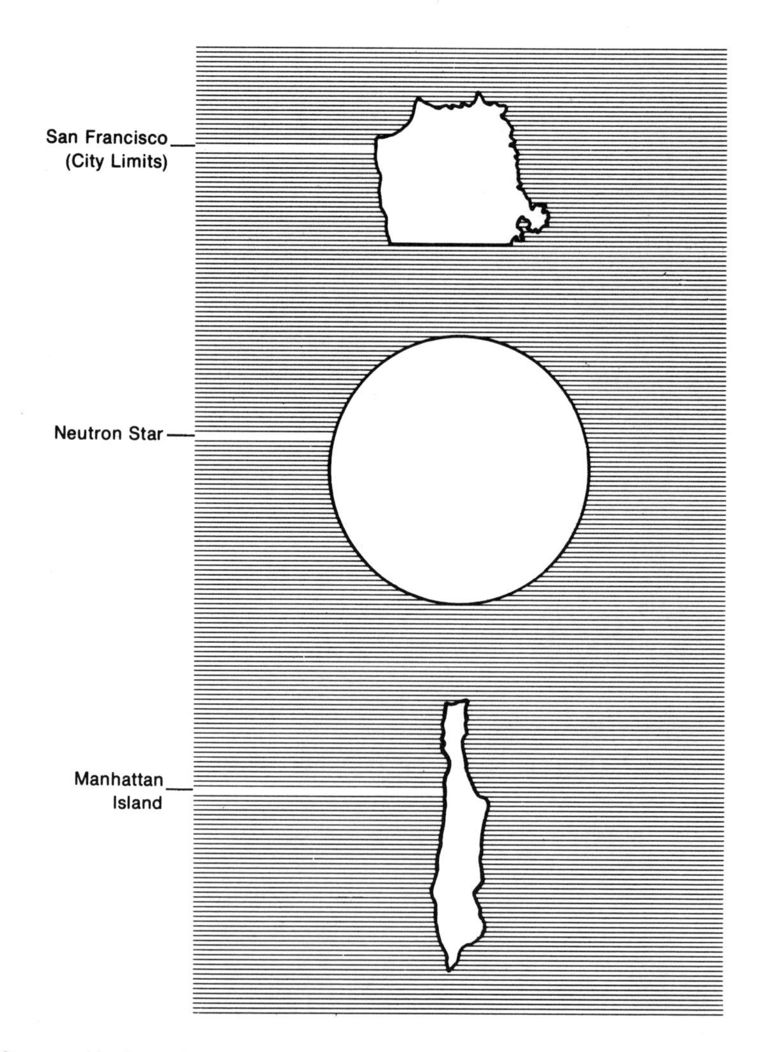

FIGURE 3-3 Neutron stars are very small.

Dutch astronomy was enduring the ordeals of Nazi occupation, Jan Oort and his friend the orientalist Duyvendak realized the importance of the "guest star" in the Chinese chronicles. Zwicky and his colleagues in California were discovering more and more supernovae and trying to unravel their story. But the neutron star idea remained just that: an idea.

In the 1960s, interest in neutron stars revived somewhat. X-ray astronomers were just beginning to send sounding rockets into the upper atmosphere to discover what x-rays were being emitted by various celestial objects. Theoreticians began trying to guess what kinds of objects might emit x-rays, and the neutron star was suggested as a possible candidate. Neutron stars, being small, would be expected to be very hot as a result of their compression. Hot objects emit high-energy radiation like x-rays, so a neutron star might be an x-ray source. This idea was stillborn, however; from what we now know about neutron stars, one of them would have to be closer than the nearest star to be detected by current x-ray telescopes if it were just shining. The seed had been planted, and neutron stars appeared once more in the literature.

PULSARS

In 1967, neutron stars were finally discovered in an unexpected way — as pulsars or pulsating radio sources. Most radio sources in the sky emit a hiss of radio noise, like static on a car radio. Pulsars, or pulsating radio sources, emit their radio-frequency radiation in regular bursts or pulses. Listen to them, and they tick each time a pulse comes through. It is now generally agreed that these pulsars are neutron stars. But how were they found?

It was at Cambridge University, England, that the appropriate equipment for discovering pulsars was set up. Jocelyn Bell and Anthony Hewish were not trying to discover pulsars. How could they — no one had thought that neutron stars would emit pulsed radio-frequency radiation. Bell and Hewish were on a different track — trying to determine the size of radio sources by watching to see whether the sources twinkled as their radio waves passed through the interplanetary medium. Just as stars twinkle and planets do not on a cold, sharp winter night, small pointlike radio sources would twinkle (or scintillate) as their radiation passed through the thin, wispy gas between the planets, while larger sources would remain steady. The scientists were able to discover the pulsars because they were looking for time variations in the strength of radio sources.

The discovery of pulsars came in the summer of 1967, as Bell noticed something rather odd on the weekly 400 feet of charts the Cambridge telescope produced. What seemed to be bursts of radio emission appeared on the records around midnight. At first, she thought that these bursts might be caused by some source of interference on the earth. One of the difficulties of radio astronomy is that many everyday objects emit radio waves: radar installations, automobile ignitions, some electric motors, snowmobile engines, and so on. You may think you have discovered something new in the sky, when all you have done is pick up the radio noise from your own refrigerator.

By the end of September, though, it became clear that the source was extraterrestrial, for it passed overhead earlier and earlier each night, just as the stars do. On November 28, the source came in very strongly, and the Cambridge astronomers were finally able to pick up the pulses. Further analysis indicated a very remarkable source: an extremely short pulse of 0.016-second's duration arrived every 1.33730115 seconds, and the pulses came in very regularly. Bell then searched through several miles of chart records, and soon three more pulsars were found. Up to this point, the Cambridge group had kept these discoveries secret, but by February 9, 1968, they were ready to announce their discovery to the world. The search now began: What were these pulsars?

The primary focus of the search was some astronomical object that would do *something* – pulsate, rotate, or finish an orbit – very regularly with periods of seconds. We needed a good pulsar clock, since the pulses from all four pulsars were evenly spaced, with accuracies of one part in ten million. (A watch that lost one second per month would be that accurate.) At one point it was thought facetiously that these pulsars might be interstellar navigation beacons for some advanced civilization, and they were jokingly dubbed LGM's (for Little Green Men). Sadly, a few tabloid newspapers got hold of this under-the-table gossip, and the *National Enquirer* blazoned forth that we astronomers had really discovered another civilization. Most definitely we had not, but such sensationalizing can occur at the time of an exciting discovery. The serious question, however, was, What are the pulsars?

The answer was not long in coming. The year 1968, following the discovery of pulsars, saw a vast amount of observational and theoretical work on these objects. Papers appeared frequently in the journals, and sometimes people were so impatient to announce their results that they used the *New York Times* to announce the discovery of a new pulsar. Yet with all this activity, it was October 1968 when the solution was discovered, at the National Radio Astronomy Observatory (NRAO) in Green Bank, West Virginia. David Staelin and Edward Reifenstein, staff members of NRAO, found a pulsar in the middle of the Crab Nebula. The Crab was known as the remnant of the 1054 supernova. The link between pulsars and some other astronomical phenomenon was forged by this discovery, since it was certain that at least one pulsar was a supernova remnant.

Theorists had been busy, too, during these hectic months. One answer to the possible nature of pulsars was the spinning neutron star idea that Thomas Gold of Cornell proposed. Others had hypothesized that the pulsars were white dwarfs or peculiar double stars. The combination of the theory that a neutron star was a supernova remnant with the presence of a pulsar in the middle of the Crab strongly supported the neutron star hypothesis. Furthermore, the only serious competitors to this idea could not explain the very rapid speed of the Crab pulsar: it pulses 30 times every second, the fastest pulsar speed. White dwarfs simply would not pulsate that fast. Numerous other arguments have now completed the matching of the neutron star models to real pulsars. Although we are still refining models and making additional observations to cement the junction between the neutron star model and the real world, the early visiion of Baade, Zwicky, Oppenheimer, and Volkoff was confirmed. Neutron stars really were the corpses left after a supernova explosion.

UNDERSTANDING PULSARS

We still do not understand how the pulses themselves are emitted. Clearly the rotation of the neutron star is the clock; every time it rotates, a pulse is emitted. The pulses originate from some kind of lighthouse mechanism; as the pulsar rotates, a beam of radiation sweeps by us (Figure 3-4). When the beam points in our direction, we see the pulse, and during the rest of the cycle, we see no pulse. A lighthouse works the same way, as a rotating shutter and lens cause the beam to sweep around the horizon, pointing in the direction of any one observer for only a short time. Some pulsars have two beams pointing in opposite directions; we see two pulses per cycle, not just one. But what produces this beam? Many people have proposed theoretical schemes, none of which has been confirmed unambiguously and all of which are possible.

What can we learn from pulsars if the pulse mechanism is not understood? The fantastic regularity of the pulses gives us an insight into the way that the neutron star rotates. The pulses are not exactly even; most pulsars have been observed to slow down and some have shown slight irregularities. Thus a pertinent beginning to examining the detailed properties of neutron stars is asking what can be learned from timing the pulsars. The pulsar clocks are slowing down very slightly; other irregularities have also been observed.

PULSAR SLOWING

The period of most pulsars is changing very slightly. The first pulsar discovered, CP 1919 or PSR 1919 + 21, increases its period by one part in 10^{15} every time it pulsates. (CP = Cambridge Pulsar, PSR =

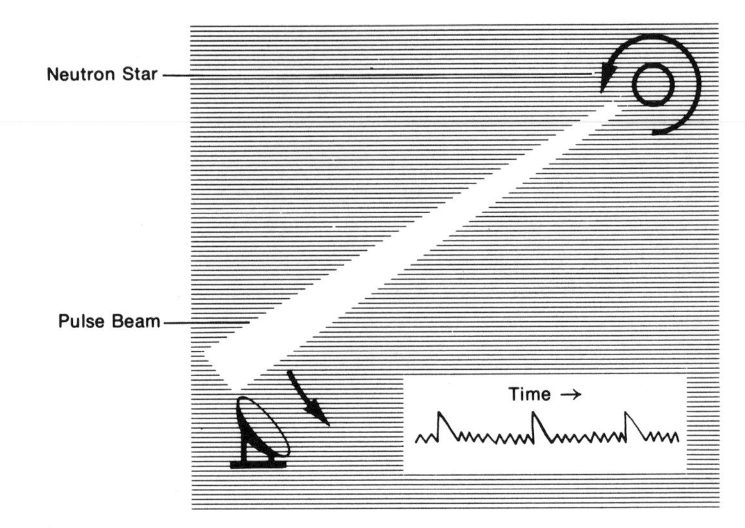

FIGURE 3-4 The pulses of pulsars come because the radio emission of pulsars is directed in a very narrow beam like the beam of a lighthouse. As the beam sweeps by the radio telescope, a sudden increase of the radio noise, and then decrease, is observed.

pulsar; the numbers refer to the pulsar's location in the sky.) One part in 10^{15} is not much; each year this pulsar's period increases by 44 billionths of a second. The reason that pulsars are slowing down is that they lose energy as they rotate. Some of this rotational energy is lost to the outside world, and as this happens, the pulsar slows down.

The slowdown of pulsars corresponds to aging. The slower pulsars are fainter and the faster ones put out more radio power. The fastest pulsar known, the one in the Crab Nebula, is one of the very brightest as well. As a result, we see no pulsars with periods longer than four seconds; their pulses are simply too faint for us to pick up. The fastest, youngest pulsars are slowing down most rapidly. The Crab pulsar will double its period in 1200 years, while CP 1919, the first pulsar discovered, will take 16 million years to double its period. Pulsars will last for tens of millions of years.

PULSAR IRREGULARITIES

Two pulsars have not conformed precisely to the pattern of a simple slowdown of the pulse period, though. In February 1969, the regular period increase of the Vela pulsar (PSR 0833–45) was being monitored weekly by the Goldstone 210-foot dish of NASA's Deep Space Network. The Goldstone astronomers noticed that at some time between

February 24 and their next weekly run on March 3 the period had decreased (Figure 3-5). What had happened?

The Crab pulsar, too, showed some irregularities. One would expect that the Crab pulsar, the youngest one, would be the least stable. The Crab has the advantage that it is producing optical pulses too; people with relatively small optical telescopes can measure the pulses and thus keep track of the neutron star clock at the center of the Crab. It is difficult to obtain observing time on larger telescopes for monitoring observations. These very precise measurements of the Crab pulsar have revealed a number of puzzling phenomena. The Crab pulsar is slowing down, but not absolutely regularly. It has speeded up, just as the Vela pulsar did, twice, in what are now called "glitches." But even with the glitches taken out, the period of the Crab pulsar seems to drift irregularly over periods of six months or so. These minor irregularities are much smaller than the regular slowdown.

Thus pulsars slow down relentlessly, speed up abruptly on occasions, and change their pulse periods very, very slightly but irregularly. What is happening to the neutron star to cause its rotation to vary in this manner? To answer this, we must first ask what a neutron star is like.

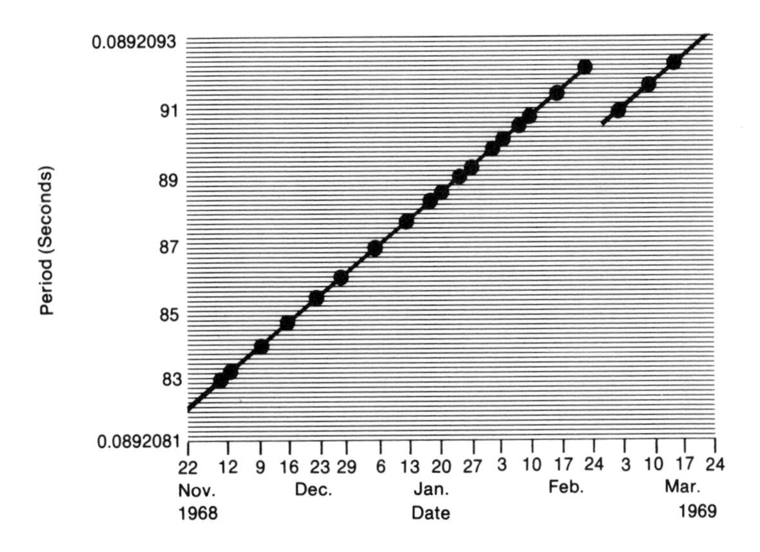

FIGURE 3-5 The Vela pulsar was slowing down in a perfectly reasonable manner until sometime between February 24 and March 3, 1969, when its period decreased by 134 nanoseconds. This speedup, observed once again in this pulsar and twice in the Crab pulsar, is called a glitch. (Adapted from *Nature* 222, p. 229, 1969.)

NEUTRON STAR STRUCTURE

While all this observational work was going on in the late 1960s and early 1970s, the theorists had renewed their attacks on the problem of the neutron star's structure. One model for the interior of a one-solar-mass neutron star is shown in Figure 3-6. The surface layer, a few meters thick, is so dominated by the pulsar's magnetic field that it may have the properties of a metal. Below this surface is a kilometer-thick layer of remarkable stuff: a crystalline solid with densities of 10^5 to 3×10^{14} g/cm^3. This solid crust is almost unbelievably dense; a thimbleful of it taken from its bottom layer would weigh as much as all the automobiles produced in the United States in a decade. This crust is solid and very hard; calculations indicate that it is 10^{17} times stiffer than steel.

Most of the neutron star consists of a superfluid core. This core holds itself up by neutron degeneracy, as the neutrons in the neutron star are resisting being compressed; they would soak up energy if they were compacted. Densities in this fluid are similar to the densities in an atomic nucleus. The nature of the central region of the neutron star is not too well understood, since the behavior of matter at densities of 10^{15} g/cm^3 and higher is poorly known. This central region may contain pions (the particles that act as nuclear glue) or hyperons (particles heavier than neutrons), or it may be a neutron solid.

We are still a long way from understanding the connection between

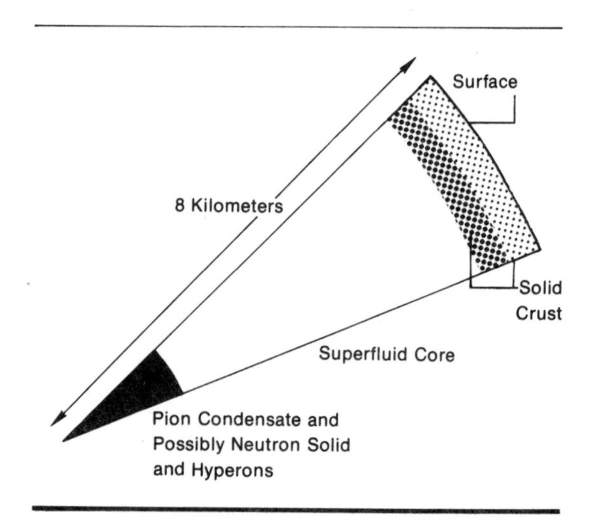

FIGURE 3-6 A slice of a neutron star. The significance of the various regions is explained in the text. (Reproduced, with permission, from "Pulsars: Structure and Dynamics," by M. Ruderman, *Annual Review of Astronomy and Astrophysics,* volume 10. Copyright © 1972 by Annual Reviews Inc. All rights reserved.)

the neutron star's interior and the irregularities in pulsar timing. It is generally thought that the irregularities come from cracking or crumbling of the crust under strain, as the pulsar readjusts its internal structure to a slower rotation rate. Many people are working on this problem, and understanding should increase in the future.

As the pulsar slows its rotation, it loses energy. Where does all this energy go? The Crab Nebula provides an answer. Most of the light from the nebula is synchrotron radiation, produced by high-energy electrons spiraling in a magnetic field. (Synchrotron radiation, a fundamental part of the quasar story, is described more completely in Chapter 8.) Adding all the energy radiated by the Crab, one can calculate that the pulsar must be supplying energy to the nebula at a rate of 4×10^{38} ergs/sec, which is quite close to the rate at which the neutron star is losing energy as it slows down. Thus the pulsar is a remarkably efficient producer of high-energy electrons.

The study of pulsars is only a few years old. To review what we know now, pulsars are neutron stars, and two have been associated with known supernova remnants. The neutron stars are rotating more and more slowly as they lose energy. The irregularities in the rotation of the neutron stars are probably connected with readjustments in the internal structure of the stars, but the nature of these readjustments is still a mystery. We might learn some more about neutron stars if we could observe them somewhere else, and there is another group of objects, recently discovered, that may give us some new insights into the neutron star.

X-RAY BINARIES

Two celestial sources of x-rays, Hercules X-1 and Centaurus X-3, have been identified with two remarkable double stars. A double star consists of two stars orbiting each other. One of these stars emits x-rays, and the x-rays heat one side of the other star so that the light from the other star varies depending on whether we are looking at the side toward the x-ray source or away from it.

The x-rays from these sources are pulsed, with periods of 1.24 seconds and 4.84 seconds, respectively, somewhat reminiscent of pulsar radio pulses. No pulses have been found in the radio range. It is now believed that the x-ray source is a stellar corpse, almost certainly a neutron star but possibly a white dwarf, that is emitting x-rays as it sucks matter away from its larger companion. This matter falls on the surface of the small star and produces x-rays as it is heated. These stars were first discovered about two years ago, and not much is yet known about them. Future analysis of them may, however, provide much useful information about the birth of neutron stars.

I have followed the neutron star story from the first musings of the 1930s about the possible existence of these things through their discovery as pulsars in 1968. We now know that these stars exist, that there is another way for a star permanently to resist the relentless force of gravity. Because neutron degeneracy pressure, like the electron degeneracy encountered in white dwarfs, has nothing to do with temperature, neutron stars can cool off without danger of collapse.

Now we have two types of stellar corpses: white dwarfs and neutron stars. Is there a third one, the black hole? Perhaps we can put white dwarfs, neutron stars, and red giants into some grand scheme for stellar evolution directed to answering the question, Do all stars become either neutron stars or white dwarfs, or do some become black holes?

EVOLUTION OF NEUTRON STARS

Figure 3-7 depicts, in a schematic way, stellar evolution. Gravity is shown constantly pulling a star to the right-hand side of the diagram, as it is always sending its orders to contract. The star can resist executing these orders by burning nuclear fuel in its center or by becoming degenerate. Nuclear fuel will sooner or later run out; it is only by becoming degenerate, by becoming a neutron star or white dwarf, that a star can avoid being pulled all the way over into the black-hole stage, the right side of the diagram.

A star at the end of the red giant stage, then, has seen its center contract repeatedly, the contraction stopping occasionally so that energy can be obtained from a nuclear reaction. Eventually the star reaches a point at which the heat generated from gravitational contraction of the core is insufficient to ignite the next fusion reaction in the sequence. How far a star proceeds in the sequence of nuclear reactions depends on its mass. A low-mass star will evolve a carbon-oxygen core by fusing helium nuclei, but its central temperatures will never reach the point at which carbon can be ignited. Intermediate-mass stars also develop a carbon-oxygen core; but when this core starts to fuse carbon and oxygen with helium to form neon, magnesium, and silicon, it does so explosively, with results that are not now known. Stars of very high mass burn carbon nonexplosively, and they probably burn their central material as far as it can go, which is to iron. No nuclear energy can be obtained from iron; energy must be added to split an iron atom or to fuse it with something else. Including additional effects such as rotation may change the details, like which stars follow which track, but it seems fairly clear that these three types of star will emerge from the red giant stage.

On the right side of the diagram we see dead stars, which have burned all their nuclear fuel and have evolved to their final state. These dead stars just cool off, not changing their structure. Two types of dead

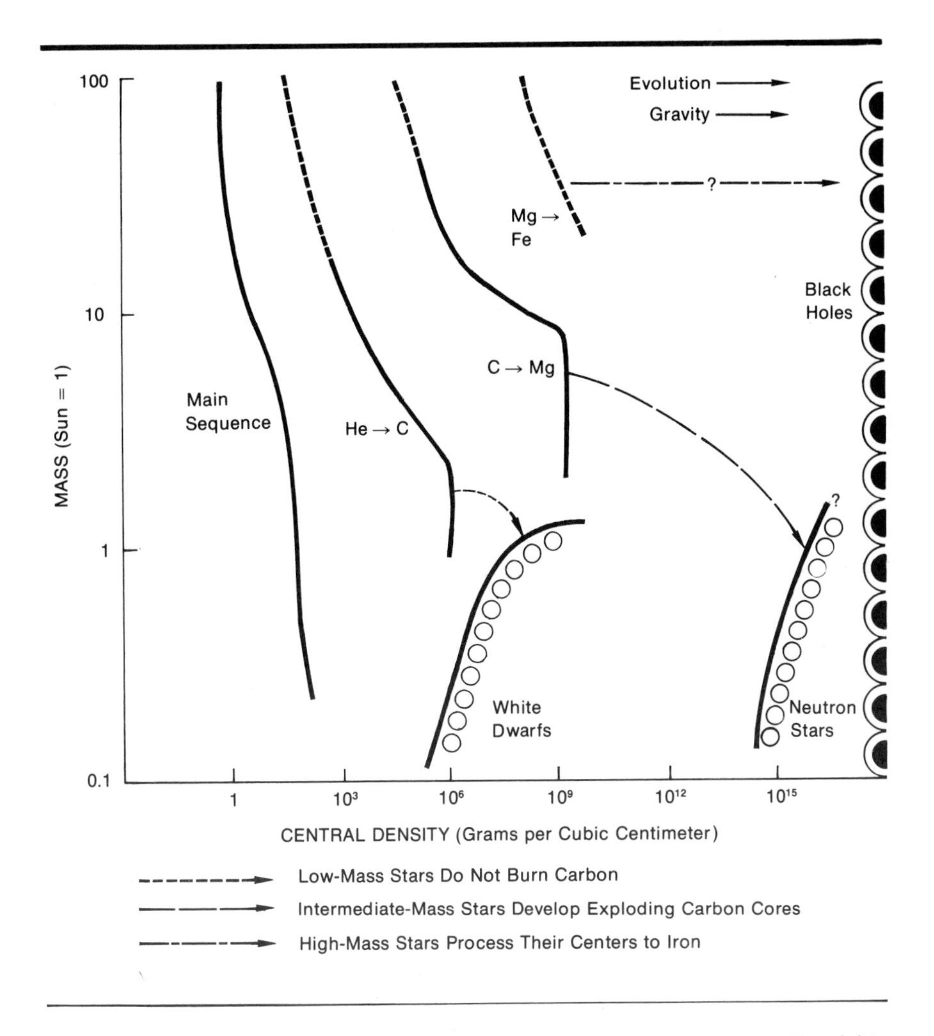

FIGURE 3-7 Outline of stellar evolution. Gravity pushes stars to the right. At the end of the red giant stage, three types of stars exist: low-mass stars, intermediate-mass stars, and high-mass stars at the end of their nuclear burning stages. (No precise data are shown for the high-mass stars.) At the extreme right are three types of stellar corpses: white dwarfs, neutron stars, and black holes, two of which are known to exist. The working hypothesis outlined in the last section of the chapter seeks to connect the three categories of stars with the three types of corpses, as shown by the dotted lines. The positions of these curves are very uncertain for stars heavier than 15 solar masses. Dashed curves are extrapolated from published theoretical models.

stars are known to exist, white dwarfs and neutron stars; and black holes are a possibility. White dwarfs and neutron stars will never collapse, no matter how much they cool off, because of degeneracy pressure. Degeneracy pressure, independent of temperature, is a barrier to collapse.

The white dwarf and neutron star barriers only work well up to a certain point, however. I previously mentioned that you could not make a white dwarf more massive than 1.4 solar masses unless you were willing to let it rotate much faster than white dwarfs are observed to do. The same phenomenon exists with the neutron stars. A neutron star that is too massive cannot hold itself up, and it must collapse to become a black hole. How big can a neutron star be? There is no neat answer to this question like the white-dwarf limit of 1.4 solar masses, since we do not know exactly how neutrons act at the extremely high densities found in the interior of neutron stars. Some investigators believe that the maximum mass of neutron stars is quite small, about 0.7 solar mass. Others, with different ideas of how neutrons interact, believe that the limiting mass is higher, about 2.2 solar masses.

There is one absolutely crucial fact that is independent of any assumptions about the way neutrons interact or anything else. No cold, nonrotating, totally evolved star – neutron star, white dwarf, or anything else – can have a mass of more than three solar masses. This superimportant finding was obtained by Princeton's Remo Ruffini and John Wheeler. Any nonrotating stellar corpse more massive than three solar masses must be a black hole. (Even if a stellar corpse is rotating, it must slow down its rotation eventually, and it will then become a black hole if it contains more mass than three solar masses.) This conclusion rests on two very firm foundations with no complications in the way. If Einstein's theory of gravitation is valid, and if causality still holds, three solar masses is the limit.

Causality refers to an assumption that is one of the basic paradigms of modern science. This assumption is: If some event, say A, *causes* some other event, say B, A must precede B from the point of view of all observers. If, for instance, I am a baseball player, and the swing of my bat causes the baseball to go over the centerfield fence in Yankee Stadium (dream!), then every person in the universe, no matter how fast he is traveling or who he is, must see the swing of my bat precede the travel of the baseball over the fence. If someone saw the ball go over the fence first, then the idea that events proceed in logical order would be overthrown. Thus, causality is a reasonable assumption. (Many science fiction stories about time travel run into problems with causality.)

The crucial question now becomes, Are there any stars that leave remnants with more mass than the magic figure of three solar masses? (The maximum mass of a neutron star is probably less than this, but I stick with the figure three because it is certain that no evolved star can have more mass than that and remain stable.) If the answer is yes, then black holes will exist, for any star that leaves a corpse of more than three solar masses leaves a black hole. We do not know the answer to this question, since the connection between red giants and dead stars is not yet firmly established. Some ideas are beginning to emerge, allowing us to draw a possible picture of the late stages of stellar evolution. This picture has now reached the status of a working hypothesis: some of us believe it, and most astronomers believe that it is a useful framework, to be proved right or wrong.

An overview of the late stages of stellar evolution

This general picture has developed from the work of many people. The seminal ideas are largely those of the Polish astronomer B. Paczynski; and J. Craig Wheeler, formerly of Harvard, has summarized the hypothesis in an article in *American Scientist* magazine.[4] Look back at Figure 3-7. We have three types of red giants: low-mass ones with inert carbon-oxygen cores, intermediate-mass ones with explosive carbon-oxygen cores, and high-mass ones, which burn all their central fuel to iron. We have three types of stellar corpses: white dwarfs, neutron stars, and black holes. The working hypothesis equates the three types of red giants to the three types of stellar corpses. Thus low-mass stars become white dwarfs, intermediate-mass stars become neutron stars, and high-mass stars become black holes. These connections are shown by the dotted lines in Figure 3-7. The divisions between low, intermediate, and high masses are not quite settled; the first division probably occurs between 2 and 6 solar masses, and the second one occurs between 8 and 20 solar masses. Let us consider the evidence about each branch of this working hypothesis. The evidence is strong for the white dwarf sequence and adequate for the neutron star sequence; and the ideas about the possible genesis of black holes are uncertain.

WHITE DWARFS

Chapter 2 discussed the evidence supporting the connection between low-mass red giants, planetary nebulae, and white dwarfs. It is almost certain that at least *some* low-mass stars become white dwarfs. They have to lose mass to do so, but the existence of planetary nebulae provides an orderly mechanism for the mass loss. This part of the scheme has much support at present.

PULSAR ORIGINS

Do pulsars, or neutron stars, form from intermediate-mass stars, or do they form from both intermediate- and high-mass stars? Or what do they form from? The existence of a pulsar in the middle of the Crab Nebula shows that in one case anyway pulsars are connected to the supernova phenomenon. The Vela pulsar is also associated with a supernova remnant. Theorists have tried to model the supernova phenomenon, and their results indicate that supernovae can form from intermediate-mass stars, as far as we can understand what is going on. Can the observations tell us anything?

CalTech astronomer James Gunn and his Princeton colleague Jeremiah Ostriker asked this question, and formulated an answer. They examined the number of pulsars in our galaxy and the distribution of these pulsars in space. The pulsar distribution is quite similar, it seems, to the distribution of intermediate-mass stars in the galaxy. Furthermore, the number of pulsars that exist is compatible with the number of intermediate-mass stars around. These findings support the idea that pulsars come from intermediate-mass stars. Subsequent work by Ostriker and coworkers reinforces this view, but there is much yet to be learned.

BLACK HOLES

Do massive stars end their lives as black holes? If black holes form, they probably form from only the most massive stars, because it is the high mass of the core that prevents a star from stabilizing itself as a neutron star or white dwarf. We may soon know observationally whether black holes exist or not; evidence now exists that they are real objects (Chapter 5). But consider the theory first.

One way to attack this question is to build a mathematical model of a massive star on a computer, project its evolution to the end, and see what the returns are. J. R. Wilson has done some calculations along this line that indicate that a massive star will form an iron core in its center. According to the computer model, the iron core will collapse, and the rest of the star will follow along, forming a black hole. His conclusion is that stars starting with more than 20 solar masses will indeed form black holes.

Wilson's calculations, however, involve some assumptions that may not be correct. Rotation is neglected; rotation might be able to halt the collapse in some way. Some explosive effects in the region immediately surrounding the iron core were neglected as unimportant or impossible to calculate. The computer can only supply the answer if we tell it the physics involved. In this case, the phenomena that were neglected so the calculations might be feasible might enable massive stars to avoid making black holes. The answer is not here yet.

If a massive star is not to become a black hole, though, it must figure out some way to expel almost all of its mass into space. The most massive star known, HD 47129, has a mass of about 76 solar masses. Its mass can be measured, as it is a member of a double star system. A neutron star, or any dead star that is not a black hole, can be no more massive than 3 solar masses. Thus if HD 47129 is to avoid becoming a black hole, it must shed most of its mass before it dies. It must get rid of at least 73 solar masses of gas, and it is not clear how this could be done. Most red giant cores contain one-third to one-half of the total mass of the star, so a massive star must lose not only its envelope but also part of its core if it is to avoid becoming a black hole.

After all of this academic hedging, you are probably thinking, "Get out of the computer center and into the real world and go find a black hole!" That is where we are headed next. First, though, we stay in the model world, as we need to know what a black hole looks like in order to know one when we see it.

White dwarfs and neutron stars exist. These stars, by virtue of degeneracy, have managed to win the battle with gravity and keep themselves from collapsing to the black hole stage. Neutron stars, whose existence was first suggested in the 1930s, were confirmed in 1967 with the discovery of pulsars. While the pulse mechanism is still unclear, the pulsar clock is a rotating neutron star. Examining the way that this pulsar clock keeps time as it rotates will, in the future, give us much information about the structure of the neutron star.

The connection between neutron stars and white dwarfs and their precursors, the red giants, is now familiar to you. An evolutionary scheme has emerged, but it needs a lot of testing. One can reasonably suppose that a third type of stellar corpse – the black hole – exists, but we need to know a bit more about stellar evolution before the theory can provide any definitive answers.

A better answer to the question, Do black holes exist? would come from the real world with the discovery of one. Let's go look. First we must know what we shall find when we see one.

JOURNEY INTO A BLACK HOLE

A black hole is the area of space surrounding an object that has shrunk to such small dimensions that its gravity becomes overwhelming. Once anything — even light — is inside the black hole, it cannot escape from the gravitational influence. Hence the name *black hole.* In principle, black holes of any size can exist, but it is difficult to see how a black hole containing less than 1.4 solar masses would form, for such small objects would form white dwarfs or neutron stars when they collapsed. A 1.4-solar-mass black hole would be five kilometers across.

But a capsule definition of a black hole does not provide enough information to allow us to discover one. We want to bring black holes from the status of theoretical objects in the model world to the status of real objects in the real world. The search for a black hole must begin with a description of what one looks like and how it interacts with the outside world. What black hole phenomena would render a black hole detectable? Exploration of the properties of a black hole also affects the world of physics, because physicists would like to use black holes as the ultimate testing ground for Einstein's theory of gravity. Holes are the only place in the universe where gravity is stronger than all other forces. Some black hole characteristics are quite odd; the theory states that there should be a singularity in the hole's center, a point at which matter is crushed to infinite density and zero volume. Does this singularity really exist? The idea sounds physically absurd. Can a theory that predicts such crazy things be correct?

The structure of black holes can be examined from two points of view. The centerpiece can be a thought-experiment, in which we follow the adventures of a courageous, suicidal, and indestructible astronaut who undertakes to explore a black hole by falling in to see what is there. One point of view is taken by the outside world, here represented by a rocket ship orbiting the hole at a safe distance. The other point of view is taken by the astronaut himself. This second approach is purely theoretical, for the astronaut could never return from the hole's interior to tell us whether our ideas are correct or not.

The standard black hole described in this chapter is a product of Einstein's theory of gravity. This black hole is one that forms after the collapse of a nonrotating star; it obeys Einstein's theory. Adding the issue of rotation, or trying to extend black hole theory to hypothetical holes that do not form from collapsing stars, or trying to modify Einstein's theory so that the singularity goes away opens several new landscapes that are not fully understood. (These frontiers will be discussed in Chapter 6.) The phenomena described in this chapter are based on well-understood theoretical results drawn from Einstein's theory of gravity.

Black holes are, so far, entirely theoretical objects. Since it is plausible to expect that they exist, their properties are worth exploring. No black hole has yet been found, although there is one object that certainly looks very much like one. Black holes exist primarily in the model world. It is very tempting, especially for people who like science fiction stories, to succumb to the Pygmalion syndrome and endow these model black holes with a reality that they do not yet possess. Beware of this.

History of the black hole idea

Pierre Simon, Marquis de Laplace, first thought of black holes in 1796. His initial musings were based on Newtonian gravity and Newton's now discredited corpuscular theory of light. Newton thought of light as little pellets or corpuscles, having properties like very small billiard balls. Laplace realized that such corpuscles could not escape from the surface of a sufficiently massive body. He wondered whether space would be full of these *corps obscurs,* as he called them. Maybe they would be as numerous as stars? However, there was no way to test his idea, and it disappeared into the libraries, never quoted or explored by others.

Shortly after Einstein's theory of gravity appeared, the German physicist Karl Schwarzschild calculated what space would look like surrounding a point-mass. He thus discovered the standard black hole as an inhabitant of the model world, but he, like Laplace before him, had no idea whether such a body could really exist. This was not determined until 1939, when J. Robert Oppenheimer and a student, Hartland Snyder, showed that a cold and sufficiently massive star must collapse indefinitely, becoming a black hole. The Oppenheimer-Snyder work, appearing about the same time as the Oppenheimer-Volkoff paper on neutron stars, reached much the same conclusion – black holes could exist. They might be real objects, not just mathematical games that people played with Einstein's theory. In the 1960s, with a revival of interest in Einstein's general theory of relativity, black holes were intensively investigated and their detailed properties were elucidated.

This history bears some resemblance to the early history of neutron stars. Both types of stellar corpse were first known as theoretical objects. Very little research was done on them until the 1960s, when advances in astronomical instrumentation and a revival of interest led them to be investigated more intensively. Unfortunately, black holes are somewhat more difficult to find than pulsars, as will become evident shortly.

Black holes and Einstein's theory of gravity are very closely tied together. You cannot describe a black hole even remotely well with Newton's theory of gravity, as Newton's theory works only when gravity is weak or speeds are small. Newton's theory may work when you try to calculate the trajectory of a thrown baseball, but close to the surface of a black hole, the effects of general relativity are overwhelming.

If black holes are so closely connected to Einstein's theory, you might well ask, What happens to them if Einstein is wrong? But Einstein's theory is almost certainly the correct theory of gravity; it is accepted by almost all working physicists. Furthermore, most rivals of Einstein's theory are really modifications of it, descriptions of gravity that differ in detail but not in spirit from Einstein's original theory of general relativity. Experiments rule out all but two alternative theories. The most serious contender of these two, the Jordan-Brans-Dicke theory, describes black holes very similar to those described by Einstein's theory. The second alternative has been proposed recently as a foil to Einstein's theory, and while no one has proved that it too makes black holes, it seems that it should, for it is just a modification of Einstein. The description of black holes in this chapter would not be changed significantly if either of these theories becomes the correct one. Furthermore, one of the objectives of black hole research is to discover whether the black hole is a reasonable object. If it is not, then Einstein's theory must be modified.

The view from a distance

One way to approach the understanding of what happens near a black hole is to suppose that you are a spaceship pilot of the future who happens to come upon one. The spaceship scenario is not necessary for our thought-experiment, but it is true that you have to be quite near a black hole to really see what is going on. Furthermore, you will have to be close enough to drop probes into it and see what happens to them when you want to explore the black hole's immediate surroundings.

You can only sense the existence of the black hole through its gravity. You cannot see a black hole. No light escapes from it – that is why it is called a black hole. The first noticeable effect sensed by a spaceship would be a weak but relentless gravitational pull. The spaceship would begin to fall toward the black hole. There would be nothing very unusual about this pull, though; gravity is ubiquitous in the universe, and any massive object would deflect the path of a spaceship.

If you wanted to explore the black hole, you might well choose to go into orbit around it. The spaceship's motion past the hole would prevent it from falling into the hole; and the ship would fall around the hole in the same way that the moon falls around the earth, following an orbit. You could measure the mass of the hole by determining exactly how much the hole was pulling you off the straight-line path you were on before you encountered the object. If you were in orbit, you could measure how long it took to complete one circuit of the hole. The bigger the hole, the faster the orbiting. If, for example, it took 3.7947 months to make a complete circle one astronomical unit away from the hole, you could deduce that the hole had ten times as much pulling power, or ten times as much mass, as the sun. (One astronomical unit is the distance from the earth to the sun, 1.495985×10^8 km.) This is just Kepler's third law, used by astronomers to deduce the masses of double stars (described in more detail in Chapter 5). Nothing new here.

It is only when you look toward the hole that you will see something a little odd. Most ten-solar-mass objects in the universe are visible. You would expect to see some sort of star at the center of your orbit; a ten-solar-mass star is generally a bright blue one. You would see nothing. You would be in orbit around an invisible object. The hole itself would be 0.08 second of arc across, or as big as a dime 15 miles away. It would take a 400-inch telescope to see the hole as a disk, even if there were anything there to see.

If you were lucky, you might be able to see the hole in another way, since you would see light rays from stars on the other side of the hole bend as they passed by the hole (Figure 4-1). You could do this only if there happened to be stars in the right places. According to Einstein, the paths of all particles, even photons, are affected by gravity. Thus the path of a photon or light ray is bent by a gravitational field in the same way that the earth bends the path of a thrown baseball and causes it to fall. Since the paths of light rays from distant stars would be bent as they passed by the hole, these stars would seem to be out of position. The light must fall around the hole to reach the spaceship. This bending of light has been observed near the sun, as a shift of star positions during a solar eclipse. Thus there is nothing very new about this effect; it is just that near a black hole the effect would be considerably larger.

Unfortunately neither of these two methods, which a spaceship pilot near a black hole could use to detect the hole's presence, would work from the surface of the earth, far away from the hole. Obviously

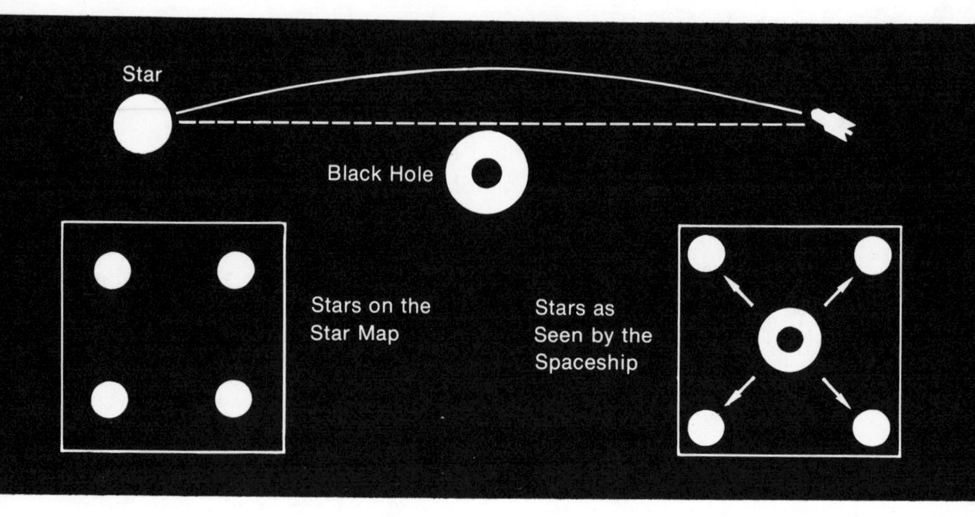

FIGURE 4-1 If you look toward a black hole from a distance, light must fall around the hole to reach your eyes. Stars near the hole would appear to be out of position.

you need to be near the hole to go into orbit around it, and the bending of light by the hole would be minuscule as seen from the earth. No equipment now available could detect it. Observations of holes must be based on phenomena arising from the interaction of holes with the material around them.

Thus our exploration of the hole will have to be extended to the depths of the hole. Furthermore, it is impossible to find out very much about the nature of a black hole simply by looking at it, since there is nothing much to see. Our spaceship will have to send a probe towards the hole and see what happens to it.

In the next few pages, for the sake of definiteness, I shall present the numerical details as they would apply to an astronaut exploring a hole of ten solar masses. Black holes forming from stellar collapse will be roughly this size. It is unlikely that any significantly smaller holes will exist in the real universe; no one has figured out a way to make a hole of less-than-stellar mass. Large holes with upwards of 10^6 solar masses may exist at the center of active galaxies and quasars as the energy source of these objects, according to one idea. Events around a large hole would be qualitatively the same as events around a smaller hole, except for the strength of the tidal forces, which would be the first phenomena encountered by the probe that was dropped into the hole.

TIDES NEAR A BLACK HOLE

As the probe approaches the black hole, nothing unusual happens for a long time. As the peculiar effects of a black hole are only evident close to the hole, this is not surprising. The first uncomfortable effect is noticed long before the neighborhood of the hole is reached, but like the bending of light, this effect is only a familiar force amplified to uncomfortable proportions, that is, tides.

Consider the effects of gravity on a person, perhaps our heroic astronaut falling to his doom, as he falls feet first towards the black hole (Figure 4-2). His legs are nearer the hole than his head, and the gravitational force pulling on his legs will be stronger than the force on his head. The difference between these two forces is the tidal gravitational force, which if unopposed will stretch the astronaut out into a long cylinder. The force arises because the closer you are to a massive object, the stronger the gravitational force is. These tides are a common feature of the interaction between two bodies, such as the moon and the earth. This tidal effect produces the ocean tides that are a familiar effect of life along the coast. They act on our bodies all the time, since our feet are nearer the center of the earth than our heads. On the earth, however, they do not present any serious threat as they are very weak. Near a black hole they are much stronger.

Another tidal force will act as a straitjacket, squeezing the astronaut's shoulders together. All parts of the astronaut fall toward the

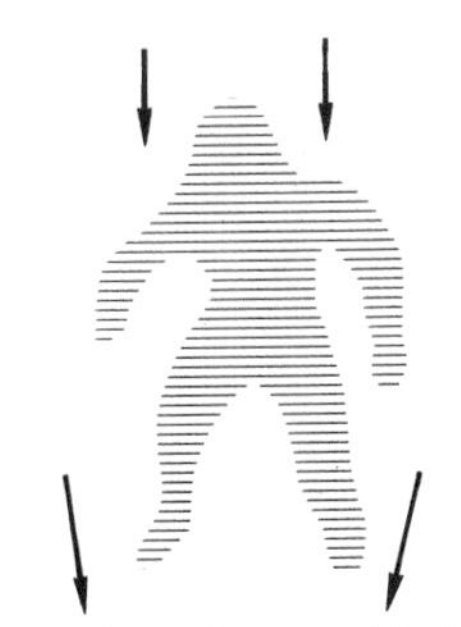

As an astronaut nears a black hole, the difference in the gravitational force on different parts of his body will exert a force on him.

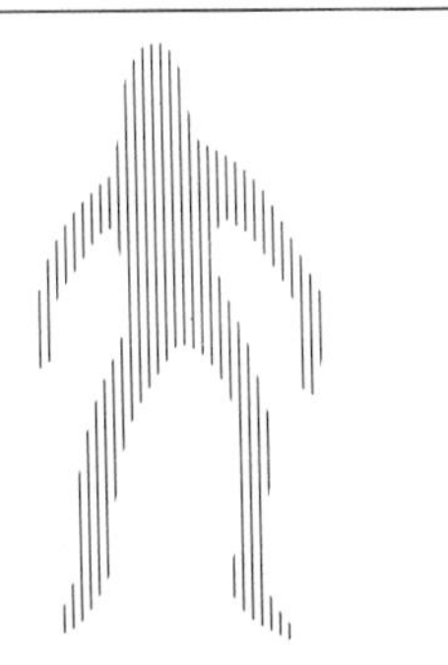

When he can no longer resist these forces, he is stretched vertically and compressed horizontally.

FIGURE 4-2 Tidal forces would distort an astronaut's body near a black hole. (The difference between the forces on different parts of the astronaut's body is exaggerated.)

center of the black hole. In particular, his two shoulders will fall on converging paths. Gravity will draw them together. It is a somewhat gruesome fate for our hero, being stretched out as though on a rack and compressed by this gravitational straitjacket. Bones and muscles must resist these forces if the body is to survive. How close can the subject get to the black hole and put up with this sort of treatment?

Optimistically, the human body can withstand a strain of ten times the earth's gravity without breaking. Our heroic astronaut would be 3000 kilometers from the ten-solar-mass black hole when the tidal forces became this strong, and he would be killed by them before he ventured any closer to the hole. It is not easy for a live astronaut to investigate the properties of a ten-solar-mass hole.

A very large black hole would be a more propitious candidate for investigation, since you can get closer to it before the tidal stresses become severe. If you were investigating a hole larger than 10^4 solar masses, you could reach the inside of the hole before the tidal forces pulled you apart. Holes this big may exist at the centers of galaxies, but as the smaller holes are likely to be more common and are certainly more detectable from the earth, I shall stick with them.

These tidal forces are the black hole phenomenon that allows us to have any chance of observing real black holes. As gas falls down towards a black hole, it is compressed by these same tidal forces that make life unpleasant for our imaginary astronaut. As this gas is compressed, it heats up. Hot gas emits high-energy radiation like x-rays, and it is these rays that are the sign of a black hole. Not all x-ray sources are black holes; you must closely investigate any black hole candidate to rule out other possible sources for these x-rays. (I shall return to this subject in Chapter 5.)

But these horrendous tides are not the phenomenon that makes the black hole one of the strangest concoctions that has been extracted from Einstein's theory of gravity. The essence of a black hole is the event horizon, the point of no return. At the event horizon, you would have to travel at the speed of light to escape from the black hole. Since no material object can travel that fast, nothing can return to the outside world once it has stepped over this invisible boundary. We explore the neighborhood of the event horizon as we watch our probe fall deeper into the hole.

Approach to the event horizon

The black hole affects space and time around it in two ways. Its gravity will distort and hinder the passage of signals from objects near it as they try to communicate with the outside world, and the passage of time near a black hole is greatly distorted. The event horizon is the edge of a black hole. Once past it, you are inside the hole, caught in its grip forever. You cannot return to the outside world. The horizon is a spherical boundary whose radius depends on the mass of the black hole. Fortunately, this radius is quite small, so the hole is small too. The radius, also called the Schwarzschild radius in honor of the discoverer of black holes, is numerically equal to 2.95 kilometers times the mass of the hole in solar masses. Our ten-solar-mass hole is thus 30 kilometers in radius or 60 kilometers across; an object this small is very difficult to see in interstellar space, much less run into. Because black holes are so small, the chances of a collision between the earth and a black hole are extremely remote.

The essence of Einstein's theory of gravity is that gravity acts on particles by distorting space and time. Thus our exploration of the ten-solar-mass hole will proceed mostly by dropping clocks into the vicinity of the hole and seeing what happens to them. The behavior of falling clocks and signals from them will be affected by the motion of the clocks themselves; for instance, the clocks will slow down because they are moving. (This is one result from Einstein's *special* relativity theory, which has nothing to do with gravity.)

Thus our thought-experiment scenario for exploring the hole will have to be a bit more complex. Our exploring probe, manned by a courageous and indestructible astronaut, will set forth on its journey into the interior with a large collection of clocks. These clocks are good clocks, reliable and accurate, and strong enough to withstand the strong tidal forces near the hole. Every once in a while, the astronaut releases a clock, tossing it into orbit around the hole. When it is in orbit, it is not moving very fast relative to the distant observer, so we can watch it from the outside and see what it is doing. At the same time, we shall observe the astronaut and see how this exploratory journey looks to us on the outside.

The events we see as we follow the black hole adventure are summarized in Table 4-1. The astronaut puts his first clock into orbit when he is 300 kilometers, or 10 Schwarzschild radii, away from the hole. What odd phenomena do we observe?

The first odd effect is that light from this clock, in orbit around the hole at a distance of 10 Schwarzschild radii, is redshifted. The photons coming from this clock will lose energy as they struggle out of the intense gravitational field near the hole. They are transformed from energetic short-wavelength photons into tired long-wavelength photons, just as a person loses energy climbing a flight of stairs, doing work against the earth's gravity. A loss of energy is a loss of frequency, or an increase in wavelength. Red photons are long-wavelength photons, and therefore this phenomenon is called a gravitational redshift.

TABLE 4-1

The Events We See as We Follow the Black Hole Adventure

DISTANCE FROM HOLE CENTER		REDSHIFT	RELATIVE CLOCK RATES	TIME	
In km	In Multiples of the Schwarzschild Radius			As Seen by the Rocket Ship	As Seen by the Probe Falling In
1 A.U.	4.96×10^{6}	0	1	0	0
300	10	0.05	1.05	204 hr 33 min 50.1129 sec	204 hr 33 min 49.6681 sec
240	8	.07	1.07	50.1135*	49.6687*
180	6	.10	1.10	50.1141*	49.6692*
120	4	.15	1.15	50.1148*	49.669666*
90	3	.22	1.22	50.1150*	49.669854*
60	2	.41	1.41	50.1153*	49.670012*
45	1.5	.73	1.73	50.1155*	49.670078*
33	1.1	2.32	3.32	50.1157*	49.670091*
30.03	1.001	30.25	31.25	50.1162*	49.670123*
$30+(3\times10^{-8288})$	$1+10^{-8289}$	$10^{4144}-1$	10^{4144}	205 hr	49.670133*
30	1	∞	∞	∞	49.670133*
15	0.5	—	—	—	49.670177*
0	0	—	—	—	49.670200*

* All items marked with the asterisks refer to 204 hr, 33 min plus the tabulated number of seconds.

In keeping with the usual notation, I list in the column labeled "Redshift" the quantity that represents the shift in wavelength per wavelength emitted by the clocks. This quantity is also noted by the letter z. Thus, if the clocks are illuminated with green light with a wavelength of 5000 angstroms, that light will be shifted by 250 angstroms as it travels to the distant rocket ship ($0.05 = 250/5000$). Such effects would be observable.

Along with the gravitational redshift, the clocks close to the hole will seem to slow down, as is shown in the column "Relative Rate." This column lists the number of seconds ticked off by the distant observer's clock in the time that it takes the clocks near the hole to tick off one second. Thus, the rocket ship's clock would tick off 1.05 seconds for every second ticked off by the clock 300 kilometers away from the hole. Clocks near the hole seem to run slowly, and events will take place in slow motion.

Once again, the gravitational redshift and the slowing of clocks are nothing very new, but close to a black hole they become extremely large. The gravitational redshift has been observed elsewhere – in white dwarfs, in the sun, and in photons sent from the basement to the top floor of Jefferson Physics Laboratory at Harvard. These two effects are related; if you run your eye down the two columns you will notice that the relative clock rate is simply $1+z$ where z is the gravitational redshift. A relation between the rates of two atomic clocks at different altitudes was one effect observed when atomic clocks were flown around the world in commercial airliners, but the principal effect observed in that experiment was not the effect of gravity. These peculiar effects near a black hole are just more dramatic versions of effects verified experimentally here on earth.

The astronaut's own clock and ours would disagree on how long it took him to reach the 300-kilometer checkpoint. He says that he deposited his first clock in orbit 204 hours, 33 minutes, and 49.6681 seconds after he left orbit. We should imagine his fall to that point to have taken a little longer, 204 hours, 33 minutes, 50.1129 seconds, a difference of 0.4448 second. Not much, it seems, just as all these other effects seem small for such an odd object as a black hole. But wait. These effects will become much larger closer to the hole.

The probe continues to fall downward, as our astronaut comes closer to the event horizon, the goal of his mission. The gravitational redshifts become larger; at 120 kilometers the redshift is 0.15, and the light illuminating the orbiting clocks looks yellow to our eyes, with a wavelength of 5600 angstroms. The orbiting clocks have slowed down in proportion, as they tick every 1.15 seconds according to our clocks, sitting in the orbiting rocket ship a safe distance from the black hole. That annoying half-second difference between our clock and the astronaut's is getting a little larger.

We must look fast at the orbiting clocks 120 kilometers away from the hole. Their orbits are unstable; they may be able to remain in orbit for a while but any deviation from a circular orbit will cause them to be captured and eventually swallowed by the hole. As the astronaut approaches the hole, putting clocks into orbit as he travels, the redshift of the light from the clocks becomes larger and larger. At 90 kilometers, the effect is truly a redshift as portrayed here. The green light illuminating the clocks will be red, with a wavelength of 6100 angstroms, by the time it reaches our eyes. At 60 kilometers, the clocks will have their light shifted beyond the red to 7000 angstroms, in the infrared part of the spectrum, by the hole's powerful gravity. We should have to look at them with an image tube, a device developed for use in Vietnam, which picks up infrared radiation. (This gadget has had many peaceful applications in astronomy, as it improves the effectiveness of telescopes.)

FIGURE 4-3 A movie showing how a distant observer would see a space probe approach a black hole. As the space probe falls in, its motion appears to freeze at the event horizon. (The numbers are meant to apply to the fall of a rocket ship toward a ten-solar-mass black hole as described in the text.)

But even the image tubes will only work well up to a point, as the redshift becomes larger and larger the closer you get to the hole. When the astronaut is 30.03 kilometers from the hole's center, or 0.03 kilometers (30 meters) from the event horizon, the supposedly green light illuminating the clocks will have a wavelength of 150,000 angstroms, far in the infrared, beyond the range of an image tube. Is there no end? Will this redshift never stop increasing? No, there is not; the redshift of photons increases without limit as you approach the event horizon.

Along with the increase in redshift comes another, more bizarre effect. Clocks near the hole are slowing down along with the redshift, as the relative clock rate is $1 + z$. Events near the hole take much more time to occur. Thirty-three kilometers away from the hole, where the redshift is 2.32, the relative clock rate is 3.32. Events this close to the hole will pass by at roughly one-third their normal rate.

Nearer and nearer the horizon, the slowdown of clocks would increase. The orbiting clocks would tick slowly, slowly, more and more slowly — tick, tick, ... At the event horizon, what would happen? It would take an infinite time until the next clock tick. Events would be frozen. Time comes to a stop at the event horizon.

What about the falling astronaut? His clock and ours are shown in Figure 4-3, which depicts graphically the events shown in the table. That half-second difference between his clock and ours would become larger and larger as he approached the event horizon. If we were monitoring his heartbeat, it would be recorded as slowing down too, along

with his clocks. He would seem to stop falling, as his fall would be frozen at the event horizon. It would be like watching a movie with someone slowing down the rate of the projector. The slowdown occurs very abruptly at the edge of the black hole; our clock would not advance to 205 hours until the astronaut was within 3×10^{-8288} km of the event horizon. (To write out 3×10^{-8288}, I should have to put 8287 zeros between the decimal point and the 3. It would take three pages in this book to write out that small a number.) His clock would be frozen, and ours would tick on. We should never see his clock go beyond 204 hours, 33 minutes, 49.670133 seconds. We should never see him fall through the event horizon. He would inch closer and closer to it, ever more slowly, but he would never pass through.

THE FROZEN STAR

If you happened to watch the formation of a black hole, you would see a somewhat similar scenario (Figure 4-4). The collapse would proceed quite rapidly at first. The light from the star would be redshifted more and more as the star came closer and closer to the horizon. (Here the redshift of the star is rendered as a darkening.) Just short of the horizon, the collapse would slow down abruptly, because the star's own gravity would cause everything to seem to happen in slow motion to a distant observer. The collapse would be effectively frozen just short of the event horizon.

Remember, it is only exceedingly close to the event horizon that the collapse appears to freeze. The large redshifts at this point would cause the star to appear to be black. The freezing of the collapse occurs only when the redshift is extremely high. After 4.6×10^{-5} sec the redshift is 10, if you start counting from the time the star has a radius of 1.5 times the

FIGURE 4-4 A movie showing the collapse of a star. As a star collapses to form a black hole, it dims very rapidly. It emits its last photon less than 0.01 seconds after it becomes smaller than 1.5 Schwarzschild radii. Compare Figure 4-3.

Schwarzschild radius. After another 4.6×10^{-5} sec, the redshift has increased tenfold again, to 100. Because the star is emitting its light in discrete photons, there is a time that the star has sent its last photon out to the outside world. Detailed calculations indicate that the last photon from a ten-solar-mass star would emerge less than 0.01 second after the star's surface passed the 1.5-Schwarzschild-radius, or 45-kilometer, point. The collapsed star would be black, and its collapse would be frozen. Hence another name for black holes: frozen stars.

THE EVENT HORIZON AS A LIMIT

The preceding section points out that the event horizon is a limit. You cannot see anything happen *at* the event horizon, as no photons can reach you from there. As you look closer and closer to the event horizon, time slows down without bound. Closer, closer, closer, the clocks will go slowly, more and more slowly. Paradoxical place, the event horizon.

The concept of the event horizon as a limit can perhaps be better illustrated by one of Zeno's paradoxes. Suppose you want to go through a door, and you are six feet away. For reasons best known to yourself, you decide to approach the door slowly, covering half of the remaining distance with each step. At first, this seems like a reasonable approach; your first step takes you three feet towards the door, and you have made progress. But you will never get through the door if you play the game according to the rules. The second step leaves you 1.5 feet away, the third 9 inches, the fourth 4.5 inches, the fifth 2.25 inches, and so on. No step will ever take you through the door, as you can only approach it with each step. The same thing happens as you look towards a black hole. If you try to watch someone enter the interior, it seems to take him longer and longer to get there as he travels more and more slowly.

Looked at from the outside, the event horizon seems to be a very strange place. Somehow the idea of time coming to a stop at the event horizon doesn't quite jibe with the way that the world is supposed to work. What sort of place is the event horizon, anyway? To explore the nature of the event horizon and the world inside it, the interior of the black hole, we shall have to succumb to the Pygmalion syndrome (recall Chapter 1) and leave the realm of the real world. Anyone who fell through the event horizon in an attempt to verify experimentally the theoretical results about to be presented could never return to tell us that we were right.

Yet there are good reasons for indulging in this theoretical exercise of imagining what a trip beyond the event horizon would be like. The idea that time comes to a stop there makes you think that maybe Einstein's theory breaks down at the event horizon. If this is true, then the very existence of black holes is open to question and the validity of Einstein's theory elsewhere in the universe is questionable. The theory is supposed to be valid anywhere in the universe, including the vicinity of the event horizon.

It turns out that the peculiarity of space-time near the event horizon — the idea of time coming to a stop — is just a consequence of our point of view. If we follow our courageous astronaut through the event horizon, we find that the horizon is not such a strange place after all. Yet I repeat that what follows is theoretical only, as no one who fell into a black hole could come out again to tell us what was really there. (Some people speculate that in fact you *could* emerge from a black hole; these speculations will be dealt with in Chapter 6.)

Through the event horizon

Look at Figure 4-3 again, and at Table 4-1, this time paying attention to the astronaut's clocks. Unlike the external observer, he will not see his fall towards the hole freeze at the event horizon. 204 hours, 33 minutes, and 49.6681 seconds after he left the rocket ship, he would be 300 kilometers away from the hole, and his clocks would be in general agreement with the clocks back on the spaceship. Only a split second later, at 49.670133 seconds, he would fall through the event horizon. As he approached the hole, he would not notice any slowing down of clocks. He would not be able to see the surface of the frozen star, for it would be black. It would look like a hole. As he fell, he would see events around him, if there were anything happening, escape from their slow-motion mode as seen from the outside and proceed normally. When he fell through the horizon, he might have to endure some discomfort from the ever increasing tidal forces. But the tidal forces would stay within bounds at the horizon, and a suitably built astronaut or probe would survive.

I cannot emphasize too strongly that at the event horizon, someone falling through would not experience any impossibly odd physical effects, there is no sign at the event horizon warning of the danger inside. No infinite tidal gravitational forces that would pull one apart before getting in. Look at the way the rocketship sees the black-hole adventure in Table 4–1. See? The probe falls through the event horizon perfectly happily, in a reasonable amount of time, from its point of view.

The absence of any pathological effects at the event horizon means that Einstein's theory does not break down there. It is only our point of view from the outside that produces the odd effect of time's coming to a stop. If you adopt another point of view, the infinite redshifts and frozen clocks disappear. They are only ephemeral, a consequence of our point of view from the outside. Einstein's theory is still valid, and it is all right to use it to predict what happens up to and through the event horizon.

WHY A BLACK HOLE?

For a long time, the term used to describe the subject of our investigation was not *black hole* but *frozen star,* or *collapsar.* We see now that the term *black hole* is quite appropriate. The appropriateness of the term *black* was discussed before; from the outside, the star would fade to invisibility in one-hundredth of a second or less if you were watching it collapse. You could not even see its surface by shining a flashlight on it, as the light from the flashlight would catch up with the collapse on its way in — the collapse would be unfrozen from the point of view of the ingoing light. It would be swallowed up by the hole. No, you cannot see the frozen star. It is black.

A similar fate would befall any foolish astronaut who sought to prove that it was not a black hole, that it was only a frozen star there in space. If he tried to scoop up a piece of the frozen star, the collapse would always unfreeze just enough to stay ahead of the scoop. If he tried too hard to pick up a piece of the star, he would pass beyond the event horizon, and disappear from his friends in the world outside. Yes, this object is truly a hole, too. The accepted term now is *black hole.* If you fall into it, you fall, and you fall, and you fall, until. . . .

THE INTERIOR OF A BLACK HOLE

There is one very serious problem with being inside the event horizon. You can never get out again once you are there. The event horizon is a cosmic turnstile. One way only. Anything, whether it is light, space probe, TV set, rock, rocking chair, or unfortunate astronaut, can go only in one direction: *inward.* (Again there are speculations that mirror images of black holes exist, in which you can go outward. Save these ideas for Chapter 6.) The one-way nature of the event horizon makes it impossible to verify experimentally what goes on inside the horizon, so we have to rely on theory as a guide.

What lies at the center of the hole? Theory will carry our probe in. (I'll call it a probe now, since the idea of a person caught inside a black hole is unduly lugubrious.) The probe is pulled relentlessly toward the center. As the probe approaches the center, the tidal forces become stronger and stronger. They increase indefinitely, so the probe will be destroyed by them before it actually reaches the center. The probe could struggle against gravity, trying to escape this fate by turning on its rocket engine and darting here and there, but it could only postpone the inevitable for a very short time. The tentacles of gravity have caught it, and it must fall to destruction at the center of the hole. In a ten-solar-mass hole, it would fall quite fast; if it did not start its engine in an attempt to escape, it would reach the center 67 millionths of a second after it passed the horizon.

What is at the center? Einstein's theory really breaks down here. The theory presents us with a very bizarre object, a singularity. A singularity is an absurdity. It is a point containing all the mass of the hole. The singularity has zero volume, and the density of matter is infinite. The tidal forces are infinite. So the theory says, anyway.

The idea that there is a singularity at the center of a black hole makes many physicists feel uncomfortable. When a theory starts producing infinities in models derived from it, a reasonable feeling develops that the theory is wrong. Standard black hole theory (the subject of this chapter) is based on Einstein's theory of gravity, so a standard black hole has a singularity in the middle. Numerous people have tried to modify Einstein's theory of gravity so that the singularity goes away.

To a certain extent, though, such modifications are beside the point. Remember the Pygmalion syndrome again. The whole purpose of this exercise of following an astronaut or probe into a hole was to see where inside the hole Einstein's theory breaks down, and in particular, whether it breaks down at the event horizon. What happens inside the event horizon has no effect on the outside world, since anything that falls in can never get out again. The interior of a black hole is cut off from our universe by the event horizon, so what happens there does not affect us.

Anything that falls into a hole loses its identity — telescope, man, or beast. All we know is that a certain amount of mass fell down the hole (Figure 4-5). Here gravity is the ultimate equalizer, since the only property of anything falling into a hole that is preserved is its mass (and its electrical charge, if it has any).

A black hole is a remarkable object. A description of it is completely intertwined with Einstein's theory of gravity. Maybe Einstein's theory is not the correct one. Can a theory that predicts these strange things possibly be correct? Recent advances in experimental techniques strongly support the idea that Einstein's theory of gravity, or some other theory very similar to it, is the correct one. Clifford Will's *Physics Today* article, listed in the Bibliography, discusses this recent work. The only two currently plausible rivals to Einstein are in a sense extensions of his theory, and the one that has been the most thoroughly investigated (the Jordan-Brans-Dicke scalar-tensor theory) makes black holes virtually identical to Einstein's.

The discovery of a real black hole would nevertheless dramatically support the general Einsteinian view on gravity. A black hole is one of the few places in the universe for which the precise nature of the theory of gravity has a significant effect. Furthermore, it is one thing to describe a theoretical black hole, and yet another to show that such things really do exist.

You can look at a black hole from two points of view — the outside and the inside. The view from the outside is the only one that can be experimentally verified. Looking at a black hole, you see an event horizon where time has come to a stop. Surrounding the event horizon, barely outside it, is the surface of the collapsing star that formed the black hole, from the collapse of this star that has been frozen. The frozen star emits no light, so it is black.

But time has not really come to a stop, as it is only your point of view, from the outside looking in, that makes you think it has. If you follow, theoretically, the adventures of a space probe dropped toward the hole, you will find out that the probe falls into the hole. Time has not stopped at the event horizon, from the point of view of someone falling in. Outsiders cannot see the collapse go to completion, but someone falling in will pass straight through the event horizon, enduring discom-

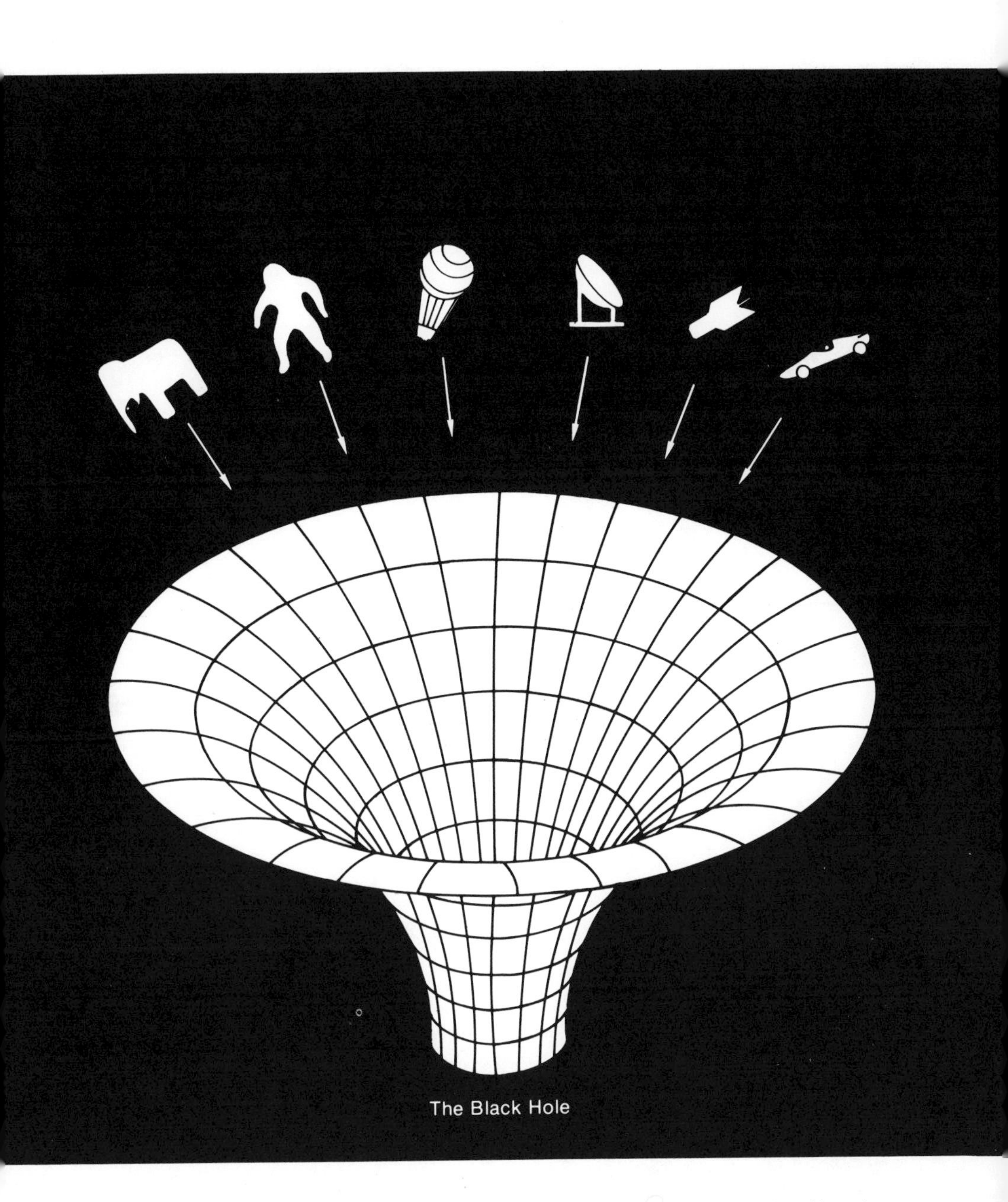

FIGURE 4-5 Everything falling into a black hole loses its identity. You do not know whether it was a space probe or a TV set that fell in. (Adapted from Remo Ruffini and John A. Wheeler, "Introducing the Black Hole," *Physics Today*, January 1970, p. 31, © American Institute of Physics.)

fort only from the tidal forces. (He would have to be pretty strong to endure those forces.) He falls through the event horizon, into the speculative arena known as the interior of the black hole. No probe, rocket, astronaut, or anything else, once inside, can escape a standard black hole. The object is caught and pulled inexorably towards the central singularity, where Einstein's theory of gravity breaks down as the forces of gravity take off towards infinity.

THE SEARCH FOR BLACK HOLES

Do black holes really exist? So far, we have examined a theoretical object, a component of the model world. Black holes are fascinating objects, but before you let yourself become too intrigued by the mysteries of the event horizon, you should try to ascertain the existence of some of these things in the real world. The search for black holes is somewhat more focused than the search for neutron stars because we have some ideas of what we should look for. In some respects, however, the stories are similar, in that it was first shown that black holes might exist (Schwarzschild, in 1916) and then shown that massive stars might actually become black holes (Oppenheimer and Snyder, in 1939); now there is a good possibility that they might be observable.

Chapters 2 and 3, which provided a sketchy overview of the life cycles of stars, reached the conclusion that massive stars may end their lives as black holes. In summary, at the end of its life, a star has burned all of its nuclear fuel. It can no longer keep its inside hot enough to hold up its surface layers, so it shrinks. If a star leaves a small corpse, the dead star can prevent itself from totally collapsing because degeneracy pressure can hold the star up. Since degeneracy pressure does not arise from heat, it will continue to operate even as the star cools. Stellar corpses with low masses thus end up as white dwarfs or neutron stars. But if a dead star is more massive than some limiting, critical mass, degeneracy pressure is overpowered by the weight of the star. Degenerate matter is simply not strong enough to hold up a massive stellar body. Although the exact value of this critical mass is not known, it is certainly less than 3 solar masses and more than 1.5 solar masses; a best current guess at this figure is somewhere near 2 solar masses. Because of the influence of factors like rotation, we cannot say that black holes must exist; we can say only that they are likely to exist.

Where are the black holes hiding?

The most logical way to look for black holes seems to be to look for dead stars with more mass than three solar masses. Such an investigation is currently going on, and the verdict is not in yet. To illustrate the techniques for the search, this chapter will investigate two cases in which black holes have been proposed as the solution to cosmic puzzles that various observations of two individual stars have created. Epsilon Aurigae is one of the strangest stars in the sky. It has been proposed that one of the components of this double system is a black hole surrounded by a vast cloud of dust. Cygnus X-1, one of the first x-ray sources to be discovered, is also a double star in which one of the components may be a black hole. The evidence here is somewhat more persuasive, as you expect to see x-rays coming from matter as it falls down a black hole.

Some people argue that by confining the search for black holes to holes that come from stars, we are limiting our scope unnecessarily. It has been proposed, for example, that the mysterious explosions at the centers of some galaxies, which will be discussed in Part Two, are caused by black holes. Unfortunately the situation here is complex; it is difficult to prove that it must be a black hole that is causing a particular galaxy to have an exceptionally bright nucleus. You must be wary of "invoking black holes," as the editors of the prestigious British journal *Nature* expressed it, as an explanation for anything that cannot be understood in a straightforward way.[1]

Two theorists invoked black holes in a widely publicized proposal that the Tunguska event of 1908 was caused by the collision of a very small black hole with the earth. On June 30, 1908, the Yenisei Valley in Siberia was devastated by a collision with some object from space — something that killed hundreds of reindeer (no humans) and scorched the forest for miles around. Because Russia was at the time preoccupied with events that had a different kind of violent impact (the Russian Revolution), the site, named Tunguska, was not visited until 1927. No remnants of the object were found.

It has been generally agreed that the object that caused this devastation was a small comet. The comet idea, while not completely worked out, explains what happened. Many details fit. The object exploded in the air, as you would expect a comet to do, since a comet is made of ice. The lack of any fragments is consistent with the comet idea. In spite of the satisfactory nature of the comet model, it was proposed that the event of 1908 involved a black hole. There are several arguments against the black hole idea — a black hole would have produced many earthquakes as it tunneled through the earth, and there is no evidence of any, and such small black holes are not expected to exist in the universe. But the strongest argument against the black hole idea is that in this case it is extraneous. To convince skeptics that black holes really do exist in the universe, we need better evidence than this. We have to find something in the sky and say, "This object can be only a black hole. Any other explanation cannot explain what we see and the black hole explanation does explain all observations."

What size of black hole should we look for? Here we have to consider plausible ways in which black holes might arise. It seems as though they would have to form from the collapse of some object. Such an origin would rule out mini–black holes, or black holes smaller than 1.4 solar masses, since objects smaller than 1.4 solar masses become white dwarfs. Mini–black holes would have to have formed in some uncalculable way in the Big Bang, the explosion that marked the beginning of the universe.

Black holes much larger than stars might also exist. A few people have proposed that black holes are the machines that power the quasars. The disadvantage of a search for massive black holes is that it is impossible to distinguish unambiguously a massive black hole from a massive energetic object. As we seek unambiguous evidence for the existence of black holes, let us turn to stars. We understand stars much better than we understand quasars.

Spectroscopic double stars

If black holes are invisible, how can we find them? The simplest effect that a hole has on its surroundings is that it causes rockets, planets, other stars, or anything nearby to fall toward it or to orbit it. If the earlier illustration of a rocket ship orbiting a black hole is changed slightly, and we let a star be substituted for the rocket ship, then we have a black hole that can be detected. You can look and see the star orbiting something, and you cannot see the object it is orbiting. Therefore it is orbiting a black hole.

This picture is nice and simple when you have a close view. Unfortunately the stars are far away, and from a distance the situation is proportionately more ambiguous. How can you detect that a star is orbiting some invisible companion if the star is tens or hundreds of parsecs away? Close analysis of the message of starlight is required. There are two main questions to be answered: How do we know that the star is orbiting in the first place? and, How do we know that the companion is truly invisible and not just a dim star hidden by the light of the star we can see?

To answer both of these questions, we must analyze the star's spectrum. Starlight, like sunlight, is a mixture of all colors of the rainbow; to photograph a star's spectrum we disperse the starlight into its different colors or wavelengths and photograph the result. Sketches of stellar spectra are shown in Figure 5-1. One can also, using varying techniques, determine the color of the star. All that we know about a star must be determined from its spectrum and its color, for the only

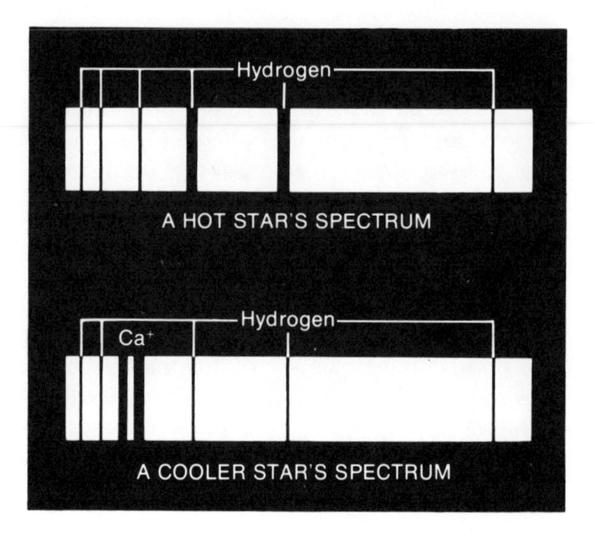

FIGURE 5-1
Stellar spectra (schematic).

messages that we receive from the star are the messages of radiation — light, radio waves, x-rays, and other forms of radiation.

The most striking feature of the stellar spectra shown in Figure 5-1 are the dark lines crossing them. These dark lines indicate that less light is being emitted at that particular color, since in the spectrum the color of the light is changing from blue on the left to red on the right. They exist because particular atoms in the surface layers of the star are absorbing light and preventing it from reaching us. (The Preliminary section explains this phenomenon in more detail.)

What can we learn from these dark lines? From their pattern, we can learn what type of star is emitting light. The spectrum can be classified, and from a star's spectral classification we can derive its temperature, size, and luminosity.

A star's temperature is the principal factor governing the appearance of its spectrum. Temperature classifications of stellar spectra are designated by letters, but the order is somewhat jumbled for historical reasons: O, B, A, F, G, K, M. For example, the bottom spectrum in Figure 5-1 is a G-type spectrum; the two very dark lines at the left come from ionized calcium atoms that are found mostly in stars with the same temperature as the sun, a G-type star. The top spectrum is that of a hotter A-type star; the calcium lines are absent.

Yet spectroscopists can tell more than just a star's temperature. In recent years, the art of spectral classification has advanced to the point that a star's size can be determined from its spectrum: supergiant, giant, or dwarf. If we can determine a star's temperature and size, we can also determine its luminosity.

We now have an idea of how to answer the second question about a star with an invisible companion. How do we know whether the invisible star is a dim companion star or a black hole — not a star at all? We determine the spectrum classification of the star that we see. From its spectral class, we can determine its luminosity. We can then determine how faint its companion must be to remain invisible. We can then ask, "Is it reasonable that the companion is so faint?"

But how do we know that the companion is even there? We are not close enough to the star–black hole pair to see the star executing a beautiful elliptical orbit as it dances around the black hole. For some pairs of stars, you can actually see the stars move around each other, but none of the black hole candidates is amenable to such an analysis. Once again, we must count on the spectrum to provide help. The exact color or wavelength of the dark lines will tell us the answer. If the wavelength of these lines changes in time — if the lines appear to shift in wavelength and then shift back again — we know that the star is alternately moving towards and away from us. If it is moving in such a fashion, it is orbiting something. Figure 5-2 shows what the observing astronomer would see.

DOPPLER SHIFTS

How does an observation like the one in Figure 5-2 tell us that there is a companion star? It is the changing color or wavelength of the dark lines in the star's spectrum that tells us that the visible star is moving in an orbit. This changing color arises from the Doppler effect.

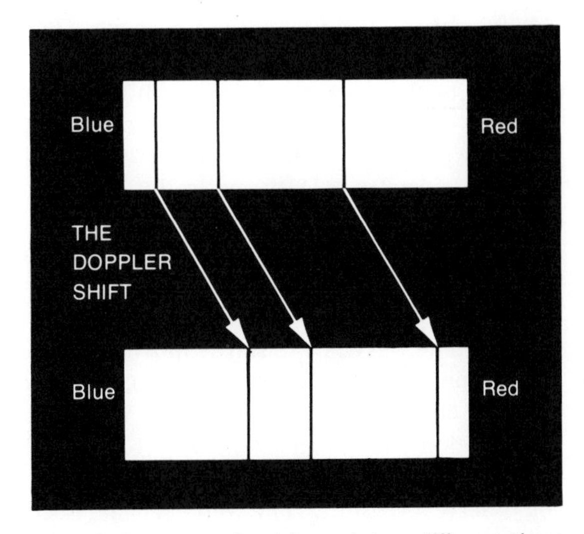

FIGURE 5-2 If the spectra of the same star taken at two different times show shifts in the wavelength (or color) of the spectrum lines, the star is alternately moving toward us and away as it orbits a companion.

Light from a star moving towards the observer looks bluer than it did when it left the star, and light moving away from the observer looks redder. For example, if two astronomers on two planets watch a star move toward one of them and away from the other, each will see a different spectrum (Figure 5-3). The astronomer on the right, seeing the star approach, will see a blueshifted spectrum and the one on the left will see a redshift. This shift connects the motion of a star to the colors or wavelengths of light in its spectrum.

What causes this Doppler shift? Consider a terrestrial analogy, shown in Figure 5-4. A police car moves down the street, like the star moving through space. This car is equipped with a siren tuned to middle C. Consider what two people on the sidewalk will observe, contrasted with what the man inside the police car hears. As the car approaches the observer on the right, sound waves from the siren will tend to pile up. The car is catching up with the sound waves it emits. As a result, the observer on the right will sense that the length of the sound waves is shrinking. He will hear a higher pitch than middle C. Similarly, the observer on the left, with the police car speeding away from him, will experience sound waves that appear stretched out. He will perceive a longer wavelength and hear a lower-pitched sound. The man inside the car does not sense that the sound waves are either stretching out or shrinking, and he will hear middle C. (The pitch of a musical sound is related to its wavelength. Short-wavelength sounds are high-pitched, and long-wavelength sounds are low-pitched.)

The extent of the Doppler shift is governed by the speed of the moving object, in this case the police car. The faster the car moves, the more the sound waves pile up (or stretch out), with consequent greater shift. In principle, a blind man with perfect pitch could determine how fast the police car was traveling. He could recognize the pitch he heard, and knowing what pitch the siren was tuned to, he could thus determine the car's velocity. (Mathematically, this relation is wavelength shift/wavelength = object velocity/wave velocity, as long as the shift is small.) You can hear the shift in pitch of a police car siren as one travels by you on the roadway. Try it sometime.

In practice, the astronomer examines spectra of a star taken at various times. When the star is moving toward the earth as it orbits its companion, the lines in the spectrum of the star will be shifted toward the blue end of the spectrum as the light waves are compressed by the star's motion. When the star passes between the earth and the companion, you observe no shift, and when the star is receding from the earth, you observe a redshift. The whole scheme is shown in Figure 5-5. (When you do work of this kind, you need to be able to measure the exact wavelength of the lines in the spectrum. The measurement is accomplished by exposing the spectrum of a fixed source on the same photographic

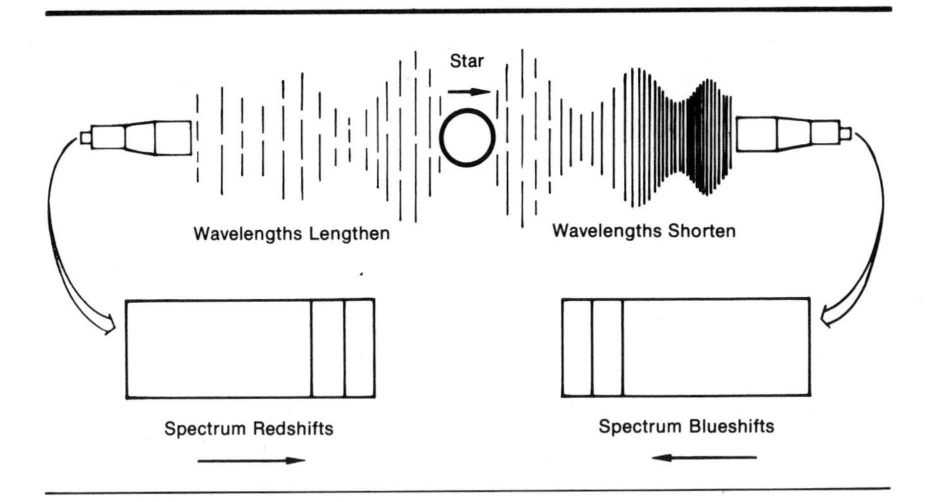

FIGURE 5-3 Astronomers see Doppler shifts in stellar spectra, as the colors (or wavelengths) of the dark lines change depending on the motion of the star. Compare with Figure 5-4.

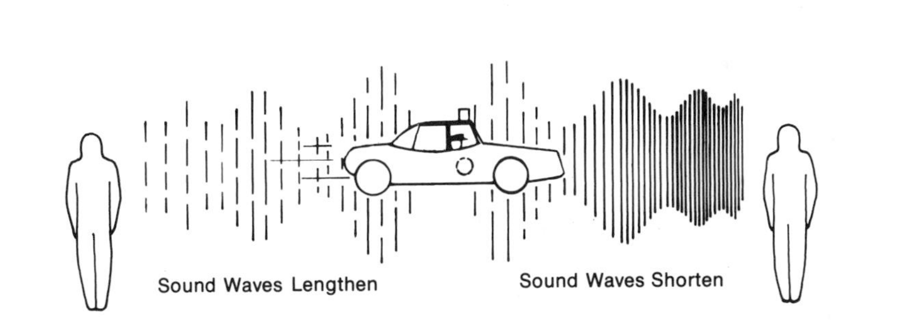

FIGURE 5-4 Pedestrians on a sidewalk can hear Doppler shifts from the siren of a passing police car.

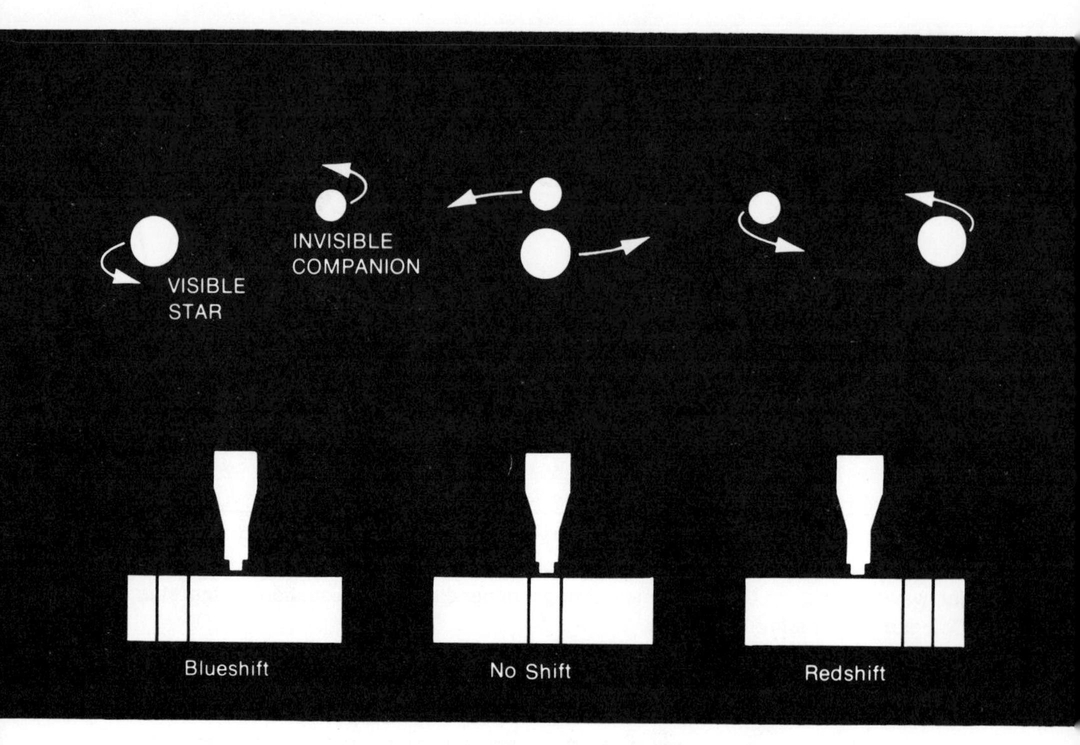

FIGURE 5-5 A single-lined spectroscopic binary, in which a star orbits a dark companion, will show a shift in its spectrum as the visible star moves toward and away from the observer in its orbit.

plate.) A star that shows changing shifts in its spectrum like those in Figure 5-5 is known as a single-lined spectroscopic binary: *single-lined,* because you observe only one spectrum; *spectroscopic,* because it takes a spectroscope to determine what is going on; and a *binary,* because the shifting pattern of spectrum lines indicates that there are two objects in the system, that the star whose spectrum you see is orbiting something.

A case history: Epsilon Aurigae

From the preceding discussion, you might be tempted to conclude that all single-lined spectroscopic binaries are black holes. Such a conclusion would be premature. All that you know about a single-lined spectroscopic binary is that the system encompasses another object or other objects. Whether that object is in fact a black hole is a further question requiring some thoughtful analysis. Each case must be considered individually, and the investigator has to ask, "Is there any way that you can explain what is observed in this system without invoking the possibility of a black hole?" For each individual star, the arguments are slightly different, but the same general questions are asked. In this section I shall take one particular example, Epsilon Aurigae, and discuss it in detail. The Epsilon Aurigae system may contain a black hole; the question is not yet decided. However, it's a star worth discussing anyway, as it is one of the strangest stars in the sky, whether or not it contains a black hole.

Auriga is a constellation that is fairly easy to recognize. It is a pentagonal group of bright stars, north of Orion, and it passes almost directly overhead in the after-dinner sky of January (see Figure 5-6). Auriga contains the bright star Capella. Just south of Capella lies a group of three stars, and Epsilon is the northernmost, brightest star of the group. It is a very peculiar spectroscopic eclipsing binary. The lines of its spectrum shift in much the same way that was described above, but in addition, every 27 years the bright star becomes dimmer as the secondary, as the companion is called, passes in front of the bright primary. All investigators agree that the secondary contains a cloud of dust that blocks out some of the light of the primary as the secondary passes between us and the primary, thus causing the eclipse. During the eclipse, and only during the eclipse, you see a second spectrum, quite similar to the spectrum of the primary star. Whether this second spectrum means that the secondary is actually emitting light of its own is unclear at present. The secondary is completely invisible outside of eclipse. Now one must attack the question, Is this invisible secondary, at the center of this immense dust cloud, a black hole?

To analyze this question, we must determine the properties of the secondary as best as we can from the information available. In particular, we need to know its mass. What do we know? We know that the primary, the star whose spectrum we see, is being pulled around by the secondary, for there are Doppler shifts in the spectrum of the primary. We can estimate the strength of the pull that the secondary is exerting. To go from the strength of this pull to the mass of the secondary, we need to know a little bit more about how double stars work.

Two stars orbiting each other are held together by gravity. The strength of the pull necessary to keep the stars from flying off in opposite directions depends on the speed of the stars in their orbit. Fast-moving stars need a strong gravitational force to hold them together, while slower-moving stars need less force. The strength of the force depends on the size of the orbit and the masses of the two stars, according to the laws of gravity. The more massive the stars, the stronger the force; and the larger the separation between the stars, the weaker the gravitational force. If we are lucky and know everything about a binary system — how

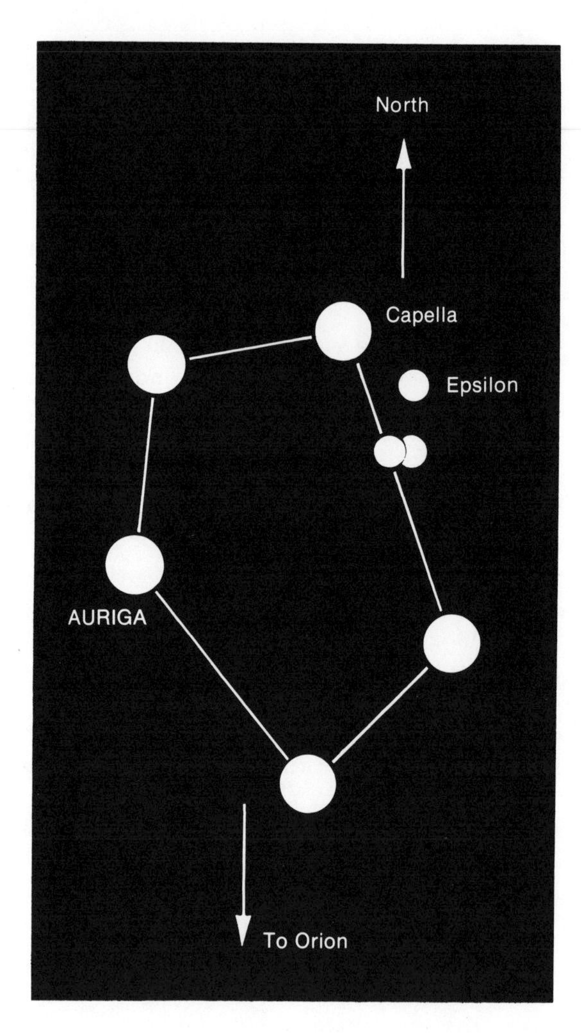

FIGURE 5-6
The constellation of Auriga and the location of the peculiar binary star Epsilon Aurigae.

fast the stars are moving and what the size of their orbit is — we can measure the masses of the two stars. There are very few stars in the sky amenable to such treatment, and Epsilon Aurigae is not one of them. It is too far away for its distance to be measured in any straightforward way and its secondary is invisible. We cannot tell whether the secondary is a small star orbiting fairly close in or a larger star somewhat farther out.

What do we do now? The art of the double-star analyst must be brought into service. You need to make assumptions about the mass of the primary and see what the known facts about the orbit can tell you. You must model the binary system, and the details of each model will differ. You examine the spectrum of the primary and see if you can estimate its properties. Its spectrum can be classified: from its spectrum you can tell that it is hotter than the sun but not as hot as Sirius. How large is it? Its spectrum looks like the spectrum of other supergiant stars in the galaxy. Such a classification indicates that it is extremely large, at least half an astronomical unit in radius. Put the center of the primary where the sun is, and Mercury would be inside the primary's surface. How massive must such a star be? Compare what the stellar evolutionists say about the evolution of massive stars with the temperature and size of the primary. Comparison shows that the primary must be a massive star, with a mass of at least 17 solar masses, with considerable uncertainty due to the uncertain nature of stellar evolution calculations. But the black hole candidate is the invisible companion or secondary. A variety of models of this system have been computed, in the interest of determining the mass of the secondary. The smallest mass that the secondary can have is eight solar masses.

Now we can dispose of one of the alternatives to the black hole hypothesis. White dwarfs and neutron stars are, like black holes, quite difficult to see. When a white dwarf is in space by itself, it is not hard to discern, but when it is right next to a normal, fairly massive star, it is overwhelmed by the glare. Thus a white dwarf or neutron star can also be an invisible companion, though not in this case, as no nonrotating white dwarf can be more massive than 1.4 solar masses and a neutron star must be less than 3 solar masses (and is probably considerably smaller than that). We can thus rule out a white dwarf or neutron star as a companion for Epsilon Aurigae. Put the details into the picture and the answer comes out: Whatever model you choose, if the secondary is a dead star, it must be a black hole.

But wait. Suppose that the secondary is a normal star that is somehow much dimmer than it should be. One thought is that the secondary is a very rapidly rotating star. Rapidly rotating stars are, in general, much dimmer than nonrotating stars of the same mass, because rotation can shoulder some of the burden of holding the star up, thus lowering the central pressure, the central temperature, and the central reaction rates. Another thought is that the dust cloud that causes the eclipses completely envelops the secondary star, obscuring it from our view. It is impossible to evaluate either of these models, in the absence of detailed calculations. You may feel that these hidden-star models are unnecessarily complex, and that it is much easier to visualize the system as a bright primary, a dust cloud, and a black hole rather than some hidden star. But the world is not always simple, and the hidden-star model is still a reasonable possibility.

With so many ambiguities, it would certainly be premature to go down to the nearest newspaper office and tell the world that we have found a black hole. (Some people did put this story in the press, and the newspapers did not stress the uncertain nature of the conclusion.)

Books and journals are a good medium for communication of incomplete results like these, as the full details of the model and assumptions behind the model can be told. The story of Epsilon Aurigae has been confined, for the most part, to the scientific literature. (The Bibliography contains references to some of the crucial papers.) Is there a black hole in Epsilon Aurigae? Whatever the final answer is, it is a very curious star, and when we have solved the riddle of this star we shall know a lot more about how stars work. It has even been proposed, for example, that the secondary contains a planetary system in formation; this is a proposal that should certainly be investigated further. But the definitive discovery of a black hole has not been made yet. Is there any way by which we can get around some of the ambiguities that the Epsilon Aurigae system puts before us? One of the newest subdisciplines of astronomy, x-ray astronomy, provides some help.

X-ray sources and black holes

The basic problem that Epsilon Aurigae presents as a black hole candidate is the difficulty of distinguishing between a black hole and a normal star embedded in a dust cloud. We could tell, from the shift of lines in the spectrum of the visible primary star, that there was another star pushing the primary around. An estimate of the strength of this pull indicated that the companion was sufficiently massive to be a black hole rather than a white dwarf or neutron star. But is the companion an evolved star, so massive that it is certainly a black hole, or is it just a normal star hidden from our eyes by this vast dust cloud? Epsilon Aurigae does not provide any evidence bearing directly on this question. The answer may emerge eventually, but only after a series of detailed models have been made. In the search for black holes, attention now focuses on evidence that can prove that the star system of interest contains an evolved star. Neutron stars and black holes are evolved stars that are so small that in the literature they are often referred to as collapsed objects, although only a black hole is, strictly speaking, collapsed.

What does a collapsed object do that stars don't do? A collapsed object is very small – some tens of kilometers across or maybe hundreds, at the utmost – while a massive star is quite large, millions of kilometers across. Remember the gruesome fate of an astronaut who ventures too close to a black hole. As the astronaut falls toward the hole, all body parts try to fall toward the same point. He is squashed by the straitjacket of gravity, as shoulders, head, feet, and the entire body fall inward toward the center. The whole astronaut is compressed. These compression forces are so strong that they would destroy any realistic space probe constructed with present-day technology that would explore a stellar-mass-sized black hole.

As gas and dust swirls toward a black hole, it will be squashed in the same way that the astronaut would be. When gas is squeezed, it heats up. The more it is compressed, the hotter it gets, and the more rapid the swirling motion is as the gas falls towards the hole. This hot vortex of infalling gas eventually gets hot and dense enough that it emits x-rays as it nears the collapsed object. Detailed theoretical models confirm this picture; if there is a black hole somewhere in space with a continuous supply of gas falling in on it, x-rays will come from the compressed, swirling gas just before it reaches the event horizon. X-ray astronomers can detect these high-energy signals that the black hole creates as it compresses the infalling gas stream.

X-RAY ASTRONOMY

X-ray astronomy, which can provide a key step in the search for black holes, is now growing up. In the 1960s, all x-ray astronomy was done with instruments sent above the atmosphere in rockets or sometimes balloons. X-rays cannot penetrate the earth's atmosphere, so it is necessary to go above the atmosphere with a rocket or satellite.

Rockets are advantageous instruments to use when you are just opening up a field and do not know whether you will find anything interesting or not. They are cheaper than satellites, both in dollars and in the amount of time it takes to put an experiment together. It takes only six months to set up a rocket experiment, if you have the instrumentation built, whereas five years often passes between the initial proposal and the launch of a satellite. Rockets have a great disadvantage in that the observing time on a typical rocket flight is only about five minutes.

The rocket era of x-ray astronomy was extremely productive, considering that the total observing time amassed by various rocket flights was little over an hour. Think how little you would know about the sky if you had just an hour to look at it. In 1969, the rocket era came to a close with the launching of the *Uhuru* satellite. This first x-ray satellite observatory has probably been one of the most productive scientific ventures of all time. (Its name comes from the time and place of its launch: off the Kenyan coast on the fifth anniversary of Kenyan independence. *Uhuru* is the Swahili word for "freedom.") Scanning the sky in the x-ray region, *Uhuru* has provided a comprehensive list of 161 x-ray sources, about 80 of which are in our galaxy. While many of these x-ray sources are *possible* black holes, it is only the compact sources, identified with dead or dying stars, that could provide unambiguous evidence of black holes.

There are three types of x-ray source that are identified with old stars. (One type, the x-ray pulsar, is mentioned in Chapter 3.) These objects, two of which have been discovered, Centaurus X-3 and Hercules X-1, are neutron stars in binary systems. The second group are the unknowns: Scorpius X-1 and Cygnus X-2. Scorpius X-1, the strongest x-ray source in the sky, is contained in a binary system with a long period and may be an old nova. The third class of compact x-ray sources contains some of the black hole candidates. These sources do not pulsate, but flicker very rapidly in intensity. Cygnus X-1 is the best known of these, but there are others. The stars X Persei and Circinus X-1 are possible black hole candidates, while 2U 0900-40, 2U 1700-37, and SMC X-1 contain x-ray sources that are not massive enough to be unambiguous black hole candidates. (The nomenclature of x-ray sources is complex. The designation *X Persei* refers to a star. The x-ray sources discovered on rocket flights were generally given the name of the constellation they were found in and a sequential number, as in Cygnus X-1. The sources found by *Uhuru* are given a catalogue number based on their position in the sky; "2U 0900-40" refers to the source in the second *Uhuru* catalogue that is at 0900-40 in the sky, or right ascension 09 hours 00 minutes, declination − 40 degrees. The designation "SMC" indicates Small Magellanic Cloud.)

The rapid flickering of some of these x-ray sources is what makes them so intriguing as black hole candidates. If an astronomical object is varying, the time scale of the variations can provide a good indication of the maximum size of the object. If an object doubles in brightness in 0.05 second, as Cygnus X-1 does, then some volume of space in the object starts to radiate in that interval. It must take no longer than 0.05 second for light to travel across that volume of space, as the entire region must receive the signal to turn on in less than 0.05 second. Whatever form that signal takes, be it an intelligent signal, a shock wave, or the actual movement of gas, the signal cannot travel faster than the speed of light. If it travels slower, the region must be even smaller. Thus the rapid variation of these x-ray sources indicates that the emitting region must be very small, very *compact* — in 0.05 second, light will travel only 15,000 kilometers, about a third of the way around the earth. The region emitting the x-rays must be smaller than this.

A black hole is a prime candidate for something smaller than the earth that produces 10^5 times the luminosity of the sun in the form of x-rays. A neutron star could, however, also produce such a high x-ray luminosity when the infalling gas hits the surface. We must investigate these systems in detail to decide whether they contain a neutron star, a black hole, or something else, also small, that is producing the x-rays.

Cygnus X-1: the first black hole?

While it is not completely certain that Cygnus X-1 is a black hole, there is good evidence that it is. What follows is a cosmic detective story. The sky has presented astronomers with a puzzle. Where does the Cygnus X-1 piece fit in? One hesitates at first to put Cygnus X-1 in the black hole slot, since invoking black holes to explain anything strange is not proving that they really exist. While there is still some debate, it is difficult to explain Cygnus X-1 as anything but a black hole.

The story begins in 1965, when Cygnus X-1 was first discovered during a rocket flight. As its name indicates, it was one of the first x-ray sources to be discovered. Its nature was uncertain, since it could not be identified with any obviously peculiar optically observed object. It is located in the Milky Way (see Figures 5-7 and 5-8), and the first x-ray positions were not precisely determined. From the initial observations, all you could say was that it was somewhere in the region shown in Figure 5-8. There are a lot of stars in that picture, and it is impossible to check all of them for peculiarities, especially when you do not know exactly how the optical counterpart to an x-ray source will be peculiar. It might even be totally invisible optically, as some x-ray sources are. In the late 1960s, more data were accumulated. The x-ray intensity of the source varied. The nature of the source was still unknown.

In 1971 and 1972, the breakthrough occurred. In March and April 1971, the *Uhuru* satellite indicated a marked change in the x-radiation from this source. Lower-energy x-rays decreased in intensity to one-quarter of their former value, and high-energy x-rays increased in intensity. But most important, a radio source appeared suddenly in the same part of the sky as the x-ray source. The 140-foot telescope of the National Radio Astronomy Observatory had been used to search for a radio counterpart to Cygnus X-1, and had not found one until this change occurred.

It seemed fairly clear that the radio source and the x-ray source were the same object. The importance of the discovery of the radio source was not that the radio noise provided much information about the nature of the source, but that the radio position could be measured much more accurately. To some extent, the added precision was unnecessary, since the x-ray positions became much more accurate at the time. (The ellipse in Figure 5-8 is the measured position of the source from an MIT rocket experiment; the experimenters said that the source was somewhere in that ellipse.) The bright star in that ellipse began to look much more intriguing, especially since the radio position was centered exactly on that star. The uncertainties in the radio position were about the same size as the image of the star in the photograph.

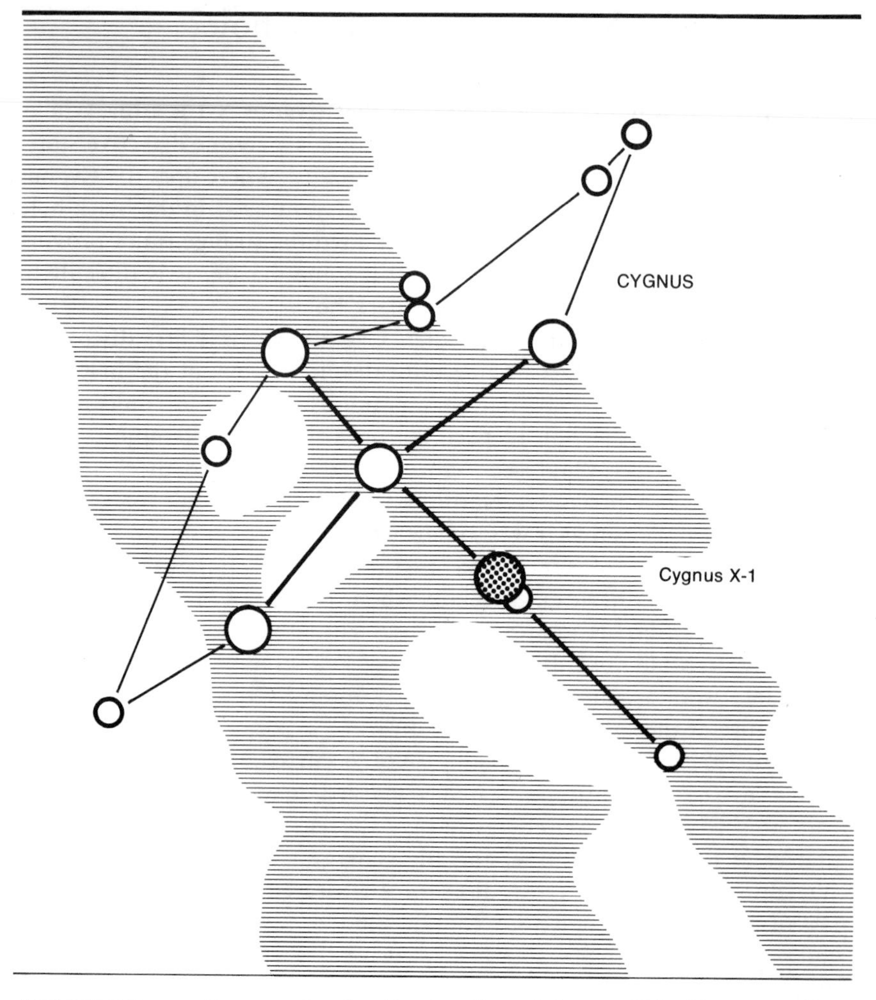

FIGURE 5-7 Cygnus, showing the location of Cygnus X-1. The heavy lines show the Northern Cross, one way of recognizing the constellation; you can make a swan out of it with the light lines. The Milky Way is also shown (lined area).

Thus the identification of Cygnus X-1 with a star was confirmed. Optical identification of x-ray sources is a critical step in determining their nature, as spectroscopy can tell you a lot about a star: how big it is, how hot it is, whether it is a double, and so forth. This star is called HDE 226868, Star No. 226,868 in the extension of the Henry Draper catalogue of spectral classifications. Its spectrum shows that it is a B-type supergiant, a large, hot, blue star.

A further discovery in 1971 was the finding by a Japanese group

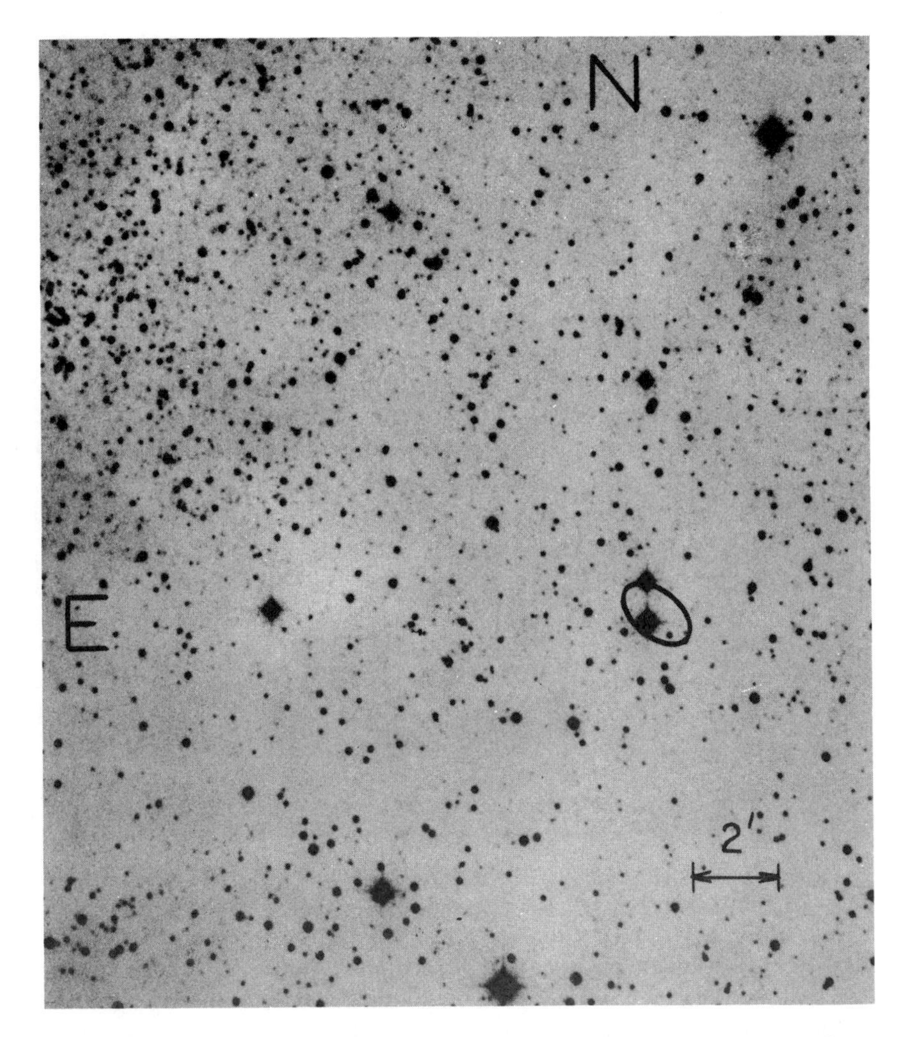

FIGURE 5-8 Enlargement of a Palomar Sky Survey print showing the location of Cygnus X-1, which is the bright star at the lower left of the ellipse. (From S. Rapaport et al., *Astrophysical Journal Letters,* vol. 168, plate L6, 1971, published by the University of Chicago Press. Copyright © by the University of Chicago. All rights reserved. The Palomar Sky Survey is copyright © by the National Geographic Society–Palomar Observatory Sky Survey.)

that the x-rays flickered very rapidly. This flickering is significant, you will recall, because it shows that the x-ray source must be very compact. Cygnus X-1 began to look more and more like a black hole.

Now the optical astronomers started working. They looked for variable Doppler shifts in the star's spectrum; these would indicate that this massive B supergiant star, HDE 226868, was being pushed around by the gravitational forces of an invisible companion. It was crucial to measure the amplitude of these Doppler shifts, and it was also necessary to determine the properties of the visible star. During the 1972 observing season (an optical astronomer can observe stars only when they are in the night sky, and Cygnus is in the night sky only in the spring and summer), Cygnus X-1 was studied intensively. From these observations, a fairly good model emerged that several investigators agreed on. While each individual adopted slightly different numbers, the basic model was the same. It is shown in Figure 5-9.

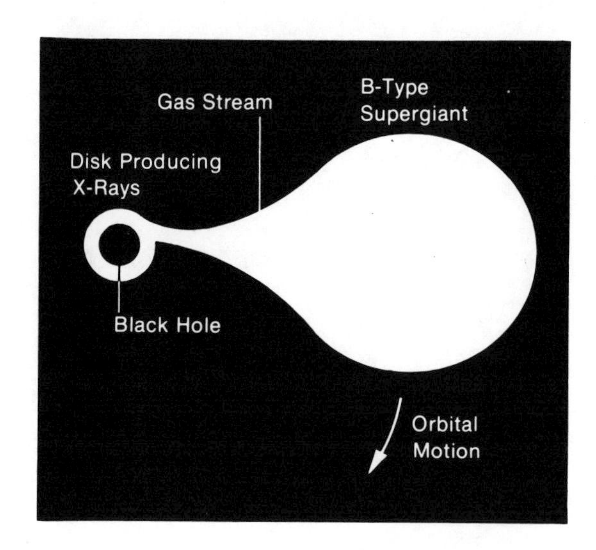

FIGURE 5-9
Model of Cygnus X-1.

The star system is a double one. The B-type supergiant is so close to the companion that its shape is distorted. The companion produces tremendous tides on the supergiant star. Furthermore, the supergiant is trying to expand further, as it moves toward the red giant stage. As it tries to expand, the companion gobbles up its outer atmosphere. Huge masses of gas flow from the surface of the B-type supergiant and form a swirling disk around the companion. Gas from this disk spirals toward the black hole, is compressed, and emits x-rays.

This gas stream shows up in other ways, too. Characteristic emission lines of hydrogen and helium come from parts of this stream. It is certain that these emission lines do not come from the star, as the Doppler shifts of these lines indicate that the gas producing them is not moving with either star.

The model for Cygnus X-1 described above and depicted in Figure 5-9 manages to explain all the phenomena. But an astronomer must be skeptical and ask, "Is there any other believable model that can explain all the observations?" The most important observation here is the existence of the x-rays, for the only reasonable way that a star can produce lots of x-rays is for the star system to contain a neutron star or a black hole. Which is it? Neutron star or black hole? We have managed to find a source for the x-rays, but have we really found a black hole? Neutron stars can have at most three solar masses. To settle the question, one must determine the mass of the companion. The binary star techniques described earlier were used to answer this question.

The general consensus at the end of 1972 was that the companion was so massive that it must be a black hole. Hot supergiants are generally massive stars, since stars that are not massive, it is believed, will be cooler and never pass through the blue supergiant stage, at least in normal stellar evolution. Typically, B-type supergiants like HDE 226868, the visible star, have masses of about 30 solar masses. With this large mass assigned to the primary, analysis indicates that the secondary must have a mass of at least five, and probably about eight, solar masses. Remember, neutron stars certainly contain no more than three solar masses, and probably considerably less than that. Therefore the companion is a black hole.

This picture was quite persuasive. Now it was time to take this news out of the scientific journals and tell the world that a black hole had been discovered. Enough evidence was in so that it was not just conjecture. But a cloud arose on the horizon.

ALTERNATIVES TO THE BLACK HOLE MODEL

This cloud on the horizon was in the form of a somewhat unlikely-looking star, HZ 22 (HZ = Humason and Zwicky; these two investigators compiled a list of faint blue stars that has turned out to be a gold mine for white dwarf seekers). Jesse Greenstein of CalTech mentioned in a conference on blue stars in 1965 that HZ 22 was an anomaly. If it was as bright as its spectrum indicated, it had no business being as far away from the galaxy as it was. It simply could not get to the outer fringes of the galaxy and still remain a blue star if it was traveling at a reasonable speed. It would have burned out long ago. But it was there. Strange star, HZ 22. But the sky is full of strange stars; you have to decide whether a particular star is likely to be strange in an interesting way before you decide to invest a great deal of time in analyzing it. Is it a single weird specimen or is it representative of a large, important group of odd fellows?

Early in 1973, HZ 22 was at last studied carefully. Greenstein discovered that it was indeed peculiar. He argued that it was not really a supergiant at all, that it was a lower-mass star, bright but not superbright, which had shed its outer envelope and had a spectrum similar to a B-type giant, although it was much less luminous than such a star. He held it to be somewhat closer than previously believed, since its faintness was due not to great distance from us but to its small luminosity. Greenstein's analysis agreed with some theoretical calculations by Virginia Trimble.

Trimble, Joseph Weber, and William Rose realized that there might be some connection between this odd blue star and Cygnus X-1. What happens to the black hole conclusion if the primary star is not very massive? Greenstein deduced that HZ 22 had a mass of only half of that of the sun. Trimble, Weber, and Rose concluded that if the primary were not a normal B-type supergiant but an analog of HZ 22, the erstwhile black hole would be small enough to be replaced by a neutron star. Not that there is anything wrong with neutron stars, but it is somewhat disappointing to see the first black hole go away.

How can you tell whether HDE 226868, the star that we observe in the Cygnus X-1 system, is a normal B-type supergiant with a black hole companion or whether it is a less massive HZ 22–type star with a neutron star companion? A crude measure of its distance will suffice. An HZ 22–type star is not as bright as a B-type supergiant. If HDE 226868 resembled HZ 22, it would be about 200 parsecs away, fairly nearby by cosmic standards. If it turned out to be a normal B supergiant, it would be about 10,000 light-years away, or about a third of the way across the galaxy. Unfortunately distances are not easy to measure in astronomy, and this distance had to be measured indirectly.

A look at Figures 5-7 and 5-8 might indicate how some clue to the distance could be obtained. Cygnus X-1 is located in the Milky Way; to see it you look in the direction of the plane of the galaxy. The Milky Way is full of interstellar clouds, vast aggregates of gas and dust. These clouds dim the light of the stars that have to shine through them. They dim blue starlight more than the red starlight, in the same way that the earth's atmosphere absorbs more blue sunlight than red sunlight at sunset. That is why the sunset is red. From the extent that HDE 226868 is reddened, or deblued, as its light passes through interstellar clouds, we can determine how much cloud stuff lies between us and it, and thus get some idea of how far away it is. The star HDE 226868 is quite red, as it resembles a yellow star rather than a blue star in color, so we can deduce that its light is passing through many interstellar clouds.

The existence of many clouds between us and HDE 226868, or Cygnus X-1, is no direct indication that the star is far away, for the Cygnus region is notorious for the extent and variety of its cloud complexes. Is there any way to make a map of the clouds that lie between the earth and this strange star?

You can examine the stars in the neighborhood of Cygnus X-1, stars that are visible as shown in Figure 5-8, and see whether they provide a clue. In the summer of 1973, two groups of astronomers worked on this problem. Careful examination of all the stars in the vicinity of Cygnus X-1 showed that all of them were behind fewer clouds and thus nearer than the black hole candidate. Their colors corresponded more closely to their temperatures, as determined from their spectral classifications. Their light was less affected by interstellar dust clouds. They must therefore be nearer than Cygnus X-1. Examination of their spectra indicated that some of these stars were intrinsically bright, and that their faintness must be due to their distance. These stars, which were nearer to us than Cygnus X-1, were about 1500 parsecs from the earth. Thus Cygnus X-1, or HDE 226868, must be at least this far away. If it were to be a star resembling HZ 22 and have a neutron star companion, it could be at most 200 parsecs away. Since such a distance is ruled out, the primary is probably a normal B-type supergiant and its companion is sufficiently massive that it must be a black hole.

CYGNUS X-1 AS A BLACK HOLE

The story is unfinished, as proposing that Cygnus X-1 is an analogue of HZ 22 is not the only way to avoid a black hole in the system. Remember that we cannot directly observe a black hole in the system; we can only infer that the black hole is there after examining the system closely. In 1974, it became difficult to avoid the presence of a black hole without invoking fairly complicated models. The fairest statement of the current situation is that Cygnus X-1 is *probably* a black hole but that it has not yet been proved that it *must* be a black hole. It is worth reviewing the steps in this somewhat complicated logical chain:

1. Cygnus X-1 is the same object as HDE 226868, a single-lined spectroscopic binary. The changes in the speed of the observable star, relative to the earth, betray the presence of a companion we cannot see. These changes show up as a changing Doppler shift in the spectrum.
2. The primary star has a normal mass for its spectral type, about 30 solar masses. Analysis of the Doppler shift indicates that the companion or companions have a mass of at least five solar masses.
3. The x-rays coming from this object indicate that the secondary is a neutron star or black hole, a collapsed object. The flickering of the x-rays indicates that the source is very small.
4. A collapsed object with a mass greater than three solar masses is a black hole. Cygnus X-1 fits this category.

Depending on how you wish to set up the links, there may be more than four steps. But if the identification of Cygnus X-1 as a black hole is to withstand the test of prolonged scientific scrutiny, each link in the chain must prove to be sound. Several questions are extant. The system may be triple, so that the companion consists of an ordinary star, with most of the mass, and a neutron star. (The past history of such a system is implausible, however, according to some calculations I have published.)

Alternatively, the companion could be a massive white dwarf, stabilized by rotation. It is not clear where the x-rays come from in the white dwarf case. It has also been proposed that the x-rays come from magnetic interactions in a more or less ordinary binary system. These ideas will be tested in the near future. Each one of us who has worked on this system has an individual opinion on the matter; I believe that Cygnus X-1 is a black hole, but that more evidence is needed to convert this belief to a certainty.

Whether these alternative models are right or wrong, they are an important contribution to the study of Cygnus X-1, as they force us to reexamine the black hole model. Scientific investigation thrives on conflict, for observers can make tests to decide between alternative models of a particular object. If you are trying merely to verify a particular model, you prove that it fits the observations in a few places, feel content, and then move on to something else. If you are trying to decide which of two models is correct, your analysis of the two models is more intensive and more directed. The model that finally turns out to be correct will then have been thoroughly checked. In addition, the investigation stimulated by the alternative models often produces some interesting new results.

Analysis of other systems similar to Cygnus X-1 will do much to illuminate the matter. More facts will probably be discovered; the lists of x-ray sources from the *Uhuru* satellite have only appeared fairly recently, and more interesting objects on that list are being found every year. It might be possible, for example, to concoct a presentable non-black-hole model for Cygnus X-1, but that model might not explain another system. Furthermore, if several systems are known, one can determine what features of the Cygnus X-1 system are common to them all. The search for black holes is only beginning.*

* The x-ray spectrum of Cygnus X-1 has been recently analyzed to see if any features of the spectrum allow one to distinguish between the black hole and neutron star models. Kip Thorne and Richard Price (*Astrophysical Journal Letters* 195 (1975), L101–L106 indicate that there is some support for the black hole model from such investigations. Perhaps another change in the character of the x-ray spectrum that occurred in May 1975 may provide more information. A recent rocket experiment (R. Rothschild, E. Boldt, S. Holt, and P. Serlemitsos, *Astrophysical Journal Letters* 189 (1974), L13–L16; E. Kellogg, *Astrophysical Journal* 197 (1975), 689–704) shows that the x-ray flux from Cygnus X-1 flickers with a time scale of milliseconds. Consequently the x-ray emitting region must be smaller than 150 km, strengthening the conclusion reached in this chapter that the source is extremely compact.

Finding black holes is a tricky business. This chapter has focused on two stars, Epsilon Aurigae and Cygnus X-1, or HDE 226868, to illuminate the issues. Since you cannot see a black hole, you can only hope to detect it through its gravitational effect on another star. This gravitational effect shows up in the changing Doppler shifts in the star's spectrum, a phenomenon that indicates that the star is orbiting an invisible companion. Detailed analysis is necessary to determine whether the companion is a black hole. To pass the black hole test, the companion must be more massive than three solar masses. X-ray emission is a great help, since x-rays probably establish the nature of the companion as a collapsed object. The verdict on Epsilon Aurigae is not in yet; it may be a black hole, and then again it may not. But it seems that it is not easy to put Cygnus X-1 anywhere in the cosmic puzzle except in the black hole slot.

FRONTIERS AND FRINGES

The black hole described in Chapter 4 is a well understood one from the point of view of the theoretician. The results presented there are accepted as fact, as long as you go along with Einstein's theory of gravity, and they are accepted by virtually all black hole theorists. The search for black holes is by no means complete; much more needs to be done. But along with the observational and astrophysical investigations involved in the search for black holes, theoreticians are currently investigating some of the more esoteric properties of black holes to see whether anything more can be learned. Such investigations are the frontiers of black hole research.

There are a few frontier areas of black hole studies that are properly called fringes, since they represent speculative ventures far beyond the boundaries of experimentally tested or even testable theory. These fringe areas are widely publicized in the news, where it is reported that black holes are space warps: you can fall into one and come out somewhere else in this universe or in another universe. While these ideas *could* be true, they are, at our present level of sophistication, flights of fancy into the never-never land inside the event horizon. It is very easy to believe that black holes are such strange objects that if you accept their existence, then anything weird, even space-warp stories, that is said about them is true. Do not fall into this trap. Black hole research, like most of science, contains some results that are true, some that are probably true, and some that are speculation — published because they are interesting if fanciful ideas and just *might* be true. I have gathered all these ideas and put them in the latter part of this chapter so that you, the reader, will know what is fact and what is not.

Frontiers of black hole research

Real stars rotate, yet the standard black hole of Chapter 4 formed from the collapse of a nonrotating spherical star. Is the rotation of the star preserved in the structure of the black hole somehow? Yes, it is, but the connection between the rotation of the star and the rotation of the black hole is not completely straightforward. In 1963, Roy Kerr produced a theoretical black hole that had rotation in it. A schematic picture of Kerr's black hole is shown in Figure 6-1.

If you approach the rotating hole from the top, it looks much like a nonrotating, Schwarzschild black hole. The event horizon is there and acts just as it did in the absence of rotation. It is only when you look at the hole from near its equator that you notice the differences. Outside the event horizon there is a boundary called the static limit. Inside the static limit, it is impossible for anyone to be at rest, as the spin of the black hole carries you around the hole, following the hole's rotation. If you wish to send signals to the outside world from the region inside the static limit, you can do so if you direct your signals in the direction of the hole's rotation. These signals will spiral around the hole and eventually get out. You can escape too, in the same manner; turn on your rocket engine and let the hole's rotation help shoot you out.

At one time it was proposed that one could extract energy from a rotating hole by falling into the ergosphere, the region between the static limit and the event horizon, turning on a rocket engine, and picking up some of the hole's rotation and carrying this rotation to the outside world. A detailed calculation has been made to find out whether one could gain any significant amount of energy from the hole by doing this, and it turned out that it was not possible. Too bad, since this had seemed to be one way to obtain enough energy to power the quasars. Yet there might be other ways in which the energy from a spinning black

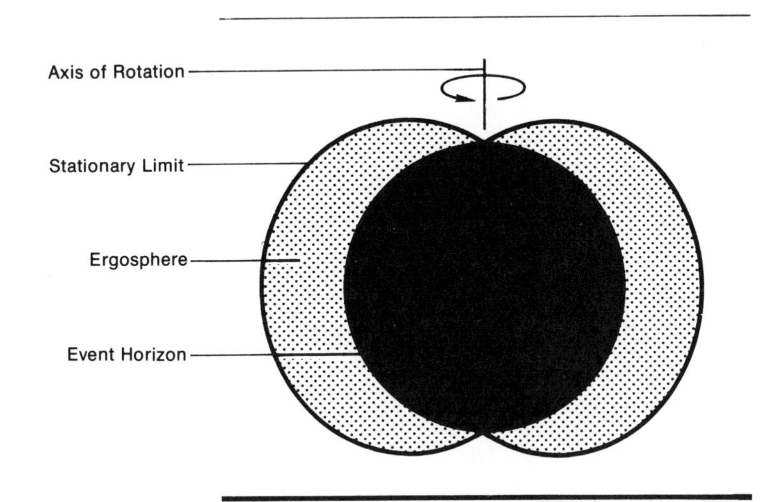

FIGURE 6-1 Cross section of a rotating black hole, showing the stationary limit, ergosphere, and event horizon.

hole could be tapped by manipulating it from the outside somehow. As will be shown in Part Two, quasars and active galaxies need a prodigious energy source to keep them going, and a massive spinning black hole in the center of a galaxy could provide that energy if some way could be found to get the energy out. Searches for such energy-extraction methods go on as part of black hole research today.

A critical aspect of the rotation problem is understanding how black holes form from rotating massive stars. Massive stars rotate rapidly. If they collapsed as solid bodies, they would be spinning too fast to form rotating black holes as we know them. Do they even form black holes at all, or do they form something else? Remo Ruffini and John Wheeler of Princeton have constructed a model in which the angular momentum will be given off by the emission of gravitational waves, but this is a model only. The transformation of a rotating star into a more slowly rotating black hole is another frontier area of black hole research.

We can create a model of a black hole that has mass and spin. It turns out that we can add electrical charge to our list of black hole properties that can be modeled, even though black holes in nature would probably be electrically neutral. Is there anything else? Are there any properties that a black hole can have besides mass, spin, and charge? One theorem that has been almost proved but not quite says that any black hole in the universe can be described by three numbers alone: its mass, its spin, and its electrical charge. If this result turns out to be true, then black holes will be quite remarkable objects, because they will not possess any individuality. Stars, for example, can be grouped into similar classes, but if you look very closely at any individual star you will find something about it that distinguishes it from its fellows. Not so the black hole. All it has are three numbers. Furthermore, if you drop something into it, all you do is to change those three numbers. The properties of the thing you dropped in are swallowed up by the hole.

Another frontier area is black hole dynamics, or what happens when two black holes encounter each other or something else. Once again, we seek to see what can be done with black holes. Do black holes just exist in space, doing nothing except when something runs into them on a very slim chance and is swallowed up? If the energy in a spinning black hole can be tapped, it could do tremendous things in the universe. Even though these areas are far from what has been observed about black holes, it is possible to conceive of ways in which some of these black hole processes can take place in nature.

These last few paragraphs have just touched on some of the ideas that black hole theorists are investigating. The primary concern is whether black holes can interact with the rest of the universe, and if so, how. Such a concern reflects the desire of black hole theorists for black holes to do something that could be observed; and it is through research avenues like these that we may be able to derive some way besides that discussed in Chapter 5 of observing black holes.

The fringe areas are concerned with what happens inside a black hole, within the event horizon. One fringe area has been widely publicized: the idea that a black hole can act like a science fiction space warp as you fall into it and come out somewhere else in the universe. These fringe areas are sheer speculation; it is difficult to see how these ideas can be tested. The frontier areas deal with the known expanses of the outer universe, outside the black hole, while the fringe areas deal with the unknown and so far unknowable lower depths of the region of space inside the event horizon.

Fringes

To seek space warps, start by looking at a map of a Schwarzschild black hole; one is shown in Figure 6-2. In this map, both space and time are plotted. Time increases toward the top of the map, and space is plotted horizontally. Such a map is called a space-time diagram. Light courses upward through this diagram along lines sloped at 45 degrees to the vertical, and slower-moving objects move along lines nearer the vertical as they cover less space in an equal amount of time than light does.

Look first at the unshaded part of the map. The heavy line at the edge of the shaded area represents the surface of the collapsing objects (star or group of stars, for instance) that formed the black hole. The event horizon is the 45-degree line in the diagram so labeled, and the singularity is drawn with black teeth.

This map is a vacuum solution to Einstein's field equations. These equations are a mathematical description of his theory of gravity. You feed into the field equations that there is some mass in space, located at the singularity, and the equations tell you how to plot a map like this. Because all the mass is concentrated at one point, and the rest of space is regarded as vacuum, the map is only valid where there really is a vacuum — away from the surface of the collapsing star. I have therefore left that part of space unshaded, to remind you that in the white section of the diagram, and only in the white section, space is empty and that the diagram is a legitimate picture of what is going on. The shaded part of the diagram is filled with the object that collapsed to form the black hole, and it is wrong to expect the vacuum solution to the field equations to be valid there.

You discover space warps by ignoring the presence of the star (or other object) that made the black hole and extending this diagram throughout all of space and time. Now you can look at the shaded region, as long as you are aware of what you are doing — neglecting the

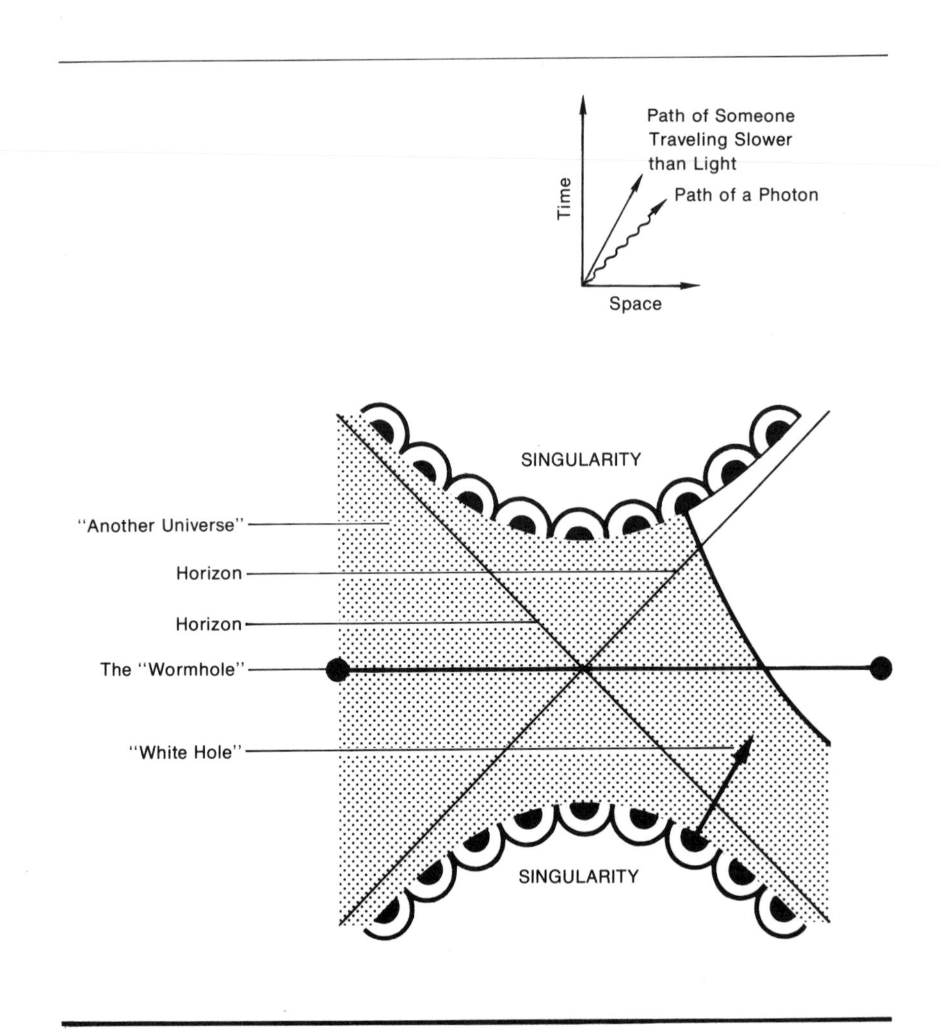

FIGURE 6-2 A map of a black hole, called a Kruskal-Szekeres map after its originators. Space is plotted horizontally and time moves generally upward. The stippled region represents the surface of the collapsing star.

star that made the black hole, or throwing away the somewhat reasonable assumption that a black hole must form from something.

If, undaunted, you look at the shaded region, it looks as though the black hole might be the gateway to another universe! You see a mirror image of the black hole on the left-hand side of the page. There are two event horizons, one located completely in the shaded region. There are two singularities, one at the bottom of the page. And it *looks* as though there are two regions of space, two universes, or two parts of the same universe, outside each horizon. The line labeled "wormhole" looks even as though it might afford some communication between the two universes. Don't take this "another universe" idea too seriously; all that we know is that in this extension of the black hole solution to the Einstein equations, we encounter a region of space with properties similar to our own, *if* (and keep this point firmly fixed in your mind as you read the rest of this chapter) *you forget about the existence of the star that made the black hole.* We do not know what this area that is labeled "another universe" is; I call it that because it is a volume of space similar to our own and is otherwise a mystery.

It seems as though you could tunnel through the path labeled "wormhole" and travel from our universe to this "other universe." In a nonrotating hole, depicted in Figure 6-2, such a trip would be impossible, since you would have to exceed the speed of light. Remember that the map is constructed so light travels along 45-degree lines; to get through the wormhole you would have to travel faster than the speed of light, which is impossible. The wormhole may, in a strictly mathematical sense, connect these two universes (remember we are forgetting the star), but it would be impossible to travel through it.

Extend these ideas to rotating black holes and you end up with a wormhole that you can travel through without exceeding the speed of light. It connects two regions of space-time that *look* like two different universes, two similar regions connected only by the black hole. But again, you can reach the other universe only by neglecting the existence of the object that formed the black hole.

Yet another weird entity can be extracted from this same point of view: the white hole. Look down at the bottom of the diagram; you see a mirror image of the top, or a time-reversed version of a black hole. It looks as though matter could follow the path labeled "white hole"—emerging from the singularity, traveling *outward* through the event horizon, and erupting into our universe.

White holes and wormholes should not be taken too seriously, however. Remember that these things came from two highly unrealistic assumptions:

1. We abstracted the ideas of black hole structure from all sources of gravity when we blithely neglected the existence of the star that made the black hole. We extended our vision into the shaded part of the diagram, and if you subscribe to the very reasonable idea (to me, at least) that a black hole must come from the collapse of a real physical object, you have no business looking at the shaded part of the diagram, and anything that you find there is a figment of your mathematical imagination.

2. We took this space on the other side of the black hole, labeled "another universe," and started thinking about it as though it *really* were

another universe. It is a mathematical entity only. Identification of this space as another universe can never be subjected to an experimental test, since something ending up in another universe will never be seen again. Those who like space warps identify that part of space with another region of *this* universe, so that you can think of somebody falling into a black hole, traveling through the wormhole (this has to be a rotating hole now), and emerging somewhere else: the "transportation system of the future." If you are to think this way, you should be aware that you are making assumptions about the nature of this part of space – a part to which you can never get anyway without running into the star that made the black hole.

The previous paragraphs are not intended to imply that wormholes will never be anything except theoretical speculation. What I mean to convey is that *at the present time* wormholes and white holes are theoretical speculation and anyone who believes that they are real objects is fostering a delusion. The status of wormholes ten or twenty years from now is anybody's guess. Today's speculation may or may not be tomorrow's fact. Black holes are at the present time much closer to reality than wormholes for two reasons:

1. We can understand, from our theoretical knowledge, how a black hole can evolve from the collapse of a known object – a massive star. It is hard to see how wormholes can form as they only are found when you take a black hole and ignore the presence of the star that formed it.
2. The existence of black holes as real objects can be verified in principle and may well have been verified in practice (Chapter 5). There is no known way to discover a worm hole, since you would have to fall into a black hole to find one, and you could never send word of your discovery back to the outside world.

Black holes and wormholes illustrate the relation among mathematical speculations, the model world, and the real world. Mathematical speculations and some work in theoretical physics deal with abstractions that may not have any real counterparts at the time at which the work is done. Some of these mathematical speculations come to be useful later, and others remain in the model world forever. At the present time, wormholes are abstractions, and it is difficult to see how wormhole research can relate to the real world of today. Consequently most black hole theorists avoid wormholes.

You deceive yourself about wormholes only if you fall victim to the Pygmalion syndrome, if you let yourself be seduced by these abstractions to the point where you think they are real. If you are not a specialist, it is very easy to read a paper on wormholes and not appreciate that it is a paper on abstractions.

Another abstract area of black hole research deals with the relation between black holes and the entire universe. Whether this is frontier or fringe depends on your point of view.

Black holes and the universe

So far, my treatment of black holes has considered the black hole phenomenon as a local one. Questions have been asked about the behavior of particles in the immediate vicinity of the black hole. Some progress has been made by investigators considering global properties of black holes, or the relations between black holes and the entire universe. Their results are difficult to assess because these "global theorems" invariably contain a few technical, seemingly innocuous assumptions about the structure of the entire universe, and these assumptions may or may not be correct.

The difference between local and global points of view can perhaps be illustrated by considering a Moebius strip. To make one, take a strip of paper, give it one twist, and glue its ends together (Figure 6-3). Compare it with another loop without a twist. In this two-dimensional analogue to our three-dimensional universe, we are like little ants crawling around the surface. Locally, the two universes look the same. An ant sitting on each of them will see the same things. But if the ant tries to walk around the loops of paper, he will end up on the other side of the loop if he lives in the Moebius-strip universe and end up back at the black hole if he lives in the ordinary-loop universe. (If you find this a bit hard to visualize, the best thing to do is cut out a piece of paper, make a Moebius strip, and play with it.) The global properties of the two loops — Moebius strip and ordinary loop — are quite different, even though their local properties are the same. Because Einstein's theory of gravitation has only been tested by local exploration, its application on a global scale is not within the domain of the experimentally tested theory.

Undaunted by these warnings, let us see what the global point of view has to tell us. I shall concentrate on the results of the global view that limit our perception of black holes, so that you can see exactly which black hole properties are based on local aspects, which are well tested according to the theory, and which come from assumptions, often technical, about global phenomena.

MUST SINGULARITIES EXIST?

The singularity (see end of Chapter 4) is a disturbing region of space, for Einstein's theory says that infinite forces exist there. The theory undoubtedly needs to be modified for the vicinity of a singularity, unless there is some way to avoid the necessity of one.

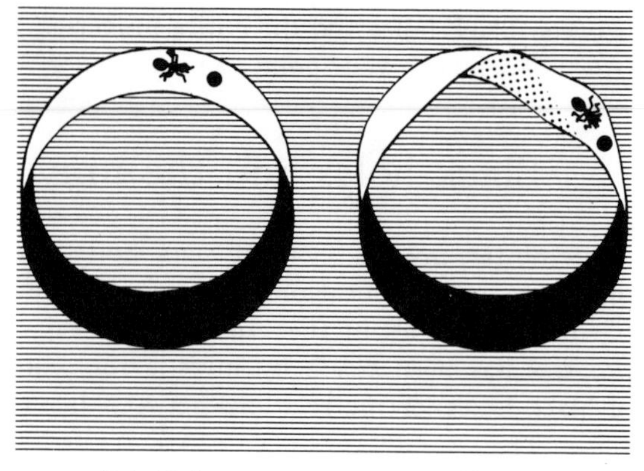

FIGURE 6-3 A paper loop and a Moebius strip, illustrating local and global view of black holes. An ant crawling around the top of the loops, near the black spot (or a black hole), will see the same phenomena, but if he crawls all the way around the Moebius strip he will end up on the other side of the loop from the black spot.

The British mathematician Roger Penrose has proved several theorems about singularities, the most general of which was the result of collaboration with Stephen Hawking and is thus known as the Penrose-Hawking theorem. It states that if an object finds itself within a trapped surface, which for our purposes can be regarded as analogous to an event horizon, a singularity must exist, provided that certain technical and apparently reasonable assumptions about the global properties of the universe hold. The existence of a singularity means that if Einstein's theory of gravity holds, the trajectories of particles that end up inside the horizon must converge at a point and end. Mathematically, it is possible that infinite forces can be avoided by ending these trajectories in another universe, whatever that means. These trajectories cannot come out somewhere else in this universe. As a result, if wormholes exist, they must connect to another universe, and other universes are beyond the domain of experimental science. Thus the Penrose-Hawking theorem provides another way of pushing wormholes, space warps, and all that stuff off into the never-never land of "other universes."

But what about this singularity? The existence of infinite gravitational forces, predicted by Einstein's theory to lie at the center of a black hole, seems absurd. Surely Einstein's theory needs to be modified. But in the standard black hole, which contains a singularity within an event horizon, it does not matter what you do to Einstein's theory inside the horizon since anything that happens inside the horizon can have no effect on the experimental world outside — no effect that can be verified by people looking at the real world. It may well be that some fantastic modification of Einstein's theory can take place near the singularity, but we shall never see it so it is irrelevant.

Is this always true, even in black holes besides the standard ones discussed here? The "hypothesis of cosmic censorship," not yet proved but probably correct, says that a singularity will always be hidden from the outside world, or clothed, by an event horizon. A singularity not inside an event horizon is called naked. A naked singularity is quite obscene, for a venturesome astronaut could go near it and subject himself to unknown forces. If all singularities are clothed in event horizons, then the breakdown of Einstein's theory near one is unimportant to the theory of black holes. We only encounter such a singularity when we consider the evolution of the entire universe (Part Three).

Our study of black holes has been on three levels. Chapter 4 considered a standard, well-understood black hole, and the results derived there are completely acceptable within the context of Einstein's theory of gravity and sufficient to inform us on what to look for when we go out to see whether black holes really exist. Chapter 6 has considered two less understood areas, the frontiers of black hole research and the fringe area phenomena. Frontier areas are far removed from the real world at present, and some people feel that they are so far removed that they need not be taken seriously at least until we have found one black hole. Nevertheless, it is at least conceivable that these frontier area discussions may eventually have some impact on the real world. Things like white holes and wormholes — all of this space-warp speculation — are in the fringe areas; such things might exist in the real world only if there are black holes that form from some process other than the collapse of a massive object and only if some unexpected new observations endow these mathematical extrapolations into "other universes" with reality.

I have spent these pages on these odd fringe phenomena so that you can tell the difference between black hole speculation and black hole theory. Black holes are such strange objects that it is very easy to accept a science-fiction-like idea as fact. It is difficult for people (myself included) to appreciate the difference between the well-grounded phenomena of Chapter 4 and the fringe-area speculations, especially when both are cloaked in the white lab coats of Science. In review, what happens near to but outside the event horizon — the idea of time coming to a stop and the tidal squashing of an astronaut — is well-founded theory. The real frontiers of black hole research, controversial areas that are nevertheless founded on a firm theoretical basis, are those areas mentioned at the beginning of this chapter: the role of rotation and their dynamics and origin. The idea of white holes and wormholes comes just from mathematical speculations of what goes on inside the event horizon, which are now fringe areas. Today's fringe can be frontier ten years from now and solid fact in another ten years, but sometimes these fringes stay fringes forever.

SUMMARY OF PART ONE

Three types of stellar corpses may exist: white dwarfs, neutron stars, and black holes. A star dies when its nuclear fires stop providing heat to maintain the star's internal pressure, which has kept the star from collapsing under the weight of its outer layers. White dwarfs and neutron stars substitute another kind of pressure for heat pressure: degeneracy pressure. Degeneracy pressure is independent of temperature, so that white dwarfs and neutron stars can cool without collapsing. White dwarfs and neutron stars are known to exist in the real world, the former as stars and the latter as pulsars.

Black holes, if they exist, would form from the collapse of very massive stars. Neutron stars and white dwarfs are certainly no more massive than three solar masses and probably a good deal smaller than that. The central feature of black holes is the event horizon, a spherical boundary that separates the inside of the hole from the outside world. The black hole phenomena that we can see from out here occur just outside of the event horizon, where something falling towards the black hole is compressed by tidal gravitational forces and appears to freeze just short of the event horizon. Some double stars exist whose invisible companions may be black holes emitting x-rays from gas compressed as it is pulled down towards the event horizon. Cygnus X-1 is probably a black hole, the evidence quite strong (though not complete) on the side of the black hole model.

The genuine frontiers of black hole research were described in Chapter 6. White holes and wormholes are interesting mathematical speculations with no foundation in the real world at present. Black hole research engenders lively controversy. The following display summarizes the results of Part One and classifies them, according to what is fact, what is informed opinion (accepted by many people but not completely proved), controversy, and speculation. This field is moving so fast that "facts" may be disproved, but I certainly believe that what is called fact will stand the test of time.

FACT	Existence of white dwarfs Neutron stars are pulsars and exist Stellar evolution through the red giant stage Theoretical model of a nonrotating Schwarzschild black hole
PROBABLE FACT	Cygnus X-1 is a black hole Pulsar irregularities are caused by starquakes Low-mass stars ⟶ Planetary nebulae ⟶ White dwarfs
WORKING MODEL	Intermediate-mass stars ⟶ Pulsars (neutron stars) Massive stars may become black holes Cosmic censorship hides all singularities behind horizons so we don't have to worry about what happens near one
CONTROVERSY	Is Epsilon Aurigae a black hole? How do pulsars produce radio emission? How massive can a neutron star be?
SPECULATION	Wormholes, white holes, and space warps

2 GALAXIES & QUASARS

Black hole studies are a theorist's paradise but an observer's hell. Models for black holes exist in profusion, but observations bearing on the subject are few. Extragalactic astronomy, the study of quasars and galaxies, is different. The observers are ahead of the theorists and observations abound, although it is not clear how these observations fit into a coherent pattern of galaxy and quasar evolution.

It was the extension of our view of the universe into other regions of the electromagnetic spectrum that led to the discovery of quasars. These objects are more luminous than galaxies, but it seems that most of their energy output originates in a volume of space only a few times larger than the solar system. Much of their emission is radio waves from fast electrons spiraling in a magnetic field (Chapter 8). Part of a quasar contains clouds of glowing gas, which produce some radiation in the visible part of the spectrum. Recently, several galaxies have been discovered that exhibit quasarlike phenomena; these active galaxies are the subject of Chapter 10. We are still far from understanding the theory of quasars; Chapter 11 discusses the ideas and speculations of the present day. There are some astronomers who believe, for varying reasons, that the quasars are not such distant objects after all. These scientists feel that the quasars are really somewhat closer, and that their redshifts come from some cause now unknown to physics. Chapter 12 discusses a controversy that has arisen about this interpretation.

As the theory of quasars is quite crude, most of the discussion in Part Two focuses on the observations. This part of the book will have a much more observational flavor than Part One did. But that flavor fits the subject; quasars were discovered by observers who were puzzling over the nature of odd-looking objects associated with certain radio sources, not by theoreticians who were attempting to explore the far reaches of the model world.

GALAXIES NEAR AND FAR

The universe is filled with galaxies; approximately one billion galaxies are within the range of earth-based telescopes. Each of these galaxies is an island universe, containing billions of stars. What do these galaxies look like? How do they evolve? How far away are they? In short, what is interesting about galaxies? We see our own galaxy, the Milky Way, at close range. This close view makes it both easier and more difficult to find out something about it.

Our galaxy, the Milky Way

On a dark night, far from the interference of city lights, a faint band of light stretches across the sky. Every civilization has had its own name for this band of light, but all names refer to it as a "way" or a "road." The Romans called it the *Via Lactea,* or as Ovid put it more poetically, the "high road paved with stars to the court of Jove."[1] The naked eye sees what looks like a band of light, but the naked-eye impression is misleading, as it is really the blurred image of countless numbers of stars.

When a small telescope or a good pair of binoculars is directed towards the Milky Way, this band of light is resolved into individual stars. We live in a disk-shaped galaxy. When your line of sight is directed in the plane of the disk, you see vast numbers of stars, or the Milky Way. When you look out of the disk, fewer stars are in your line of sight. Thus the Milky Way is the galaxy itself, and that misty band of light marks the plane of the galactic disk. A composite illustration of the Milky Way is shown in Figure 7-1. The concentration of stars towards the plane of the galaxy is evident.

If we try to map our galaxy using optical telescopes, we are hindered by the dark dust lanes, plainly visible in the Milky Way illustration. As a result, our optical view of the galaxy is a view of only a very small part of it. The eye can penetrate no further than 1000 parsecs in the plane of the galaxy, and the telescope cannot penetrate too much farther because of the absorbing interstellar dust.

FIGURE 7-1 The Milky Way as drawn in a chart with the galactic equator at the center. (Chart by Martin and Tatiana Keskula, under the direction of Knut Lundmark, Lund Observatory, Sweden.)

Dust does not absorb radio waves, however, so radio astronomers can perceive the entire galaxy and have been trying to map it. While their work is not yet complete, they have sketched a tentative map, which is shown in Figure 7-2. Two views of the Milky Way are shown, a top view and a side view. The solid parts refer to areas that are fairly well mapped, while the shaded parts represent areas of the galaxy about which our knowledge of the structure is much less certain.

The Milky Way is a disk galaxy, about 15 kiloparsecs in radius and 100 parsecs, 0.1 kiloparsec thick. We are on the outer edge, in the thin disk and about 10 kiloparsecs from the center. The disk is composed of spiral arms, which are a characteristic feature of other spiral galaxies (Figure 7-3). The only part of the spiral pattern that we are sure of comprises the spiral arms in the vicinity of the sun; we are not sure how the pattern is put together. Is the Orion arm an extension of the Cygnus arm, or does the Cygnus arm connect with the Carina arm, with the Orion "arm" just an offside spur? Our data on the more distant arms are much more scanty. The shaded arms that I have drawn represent a

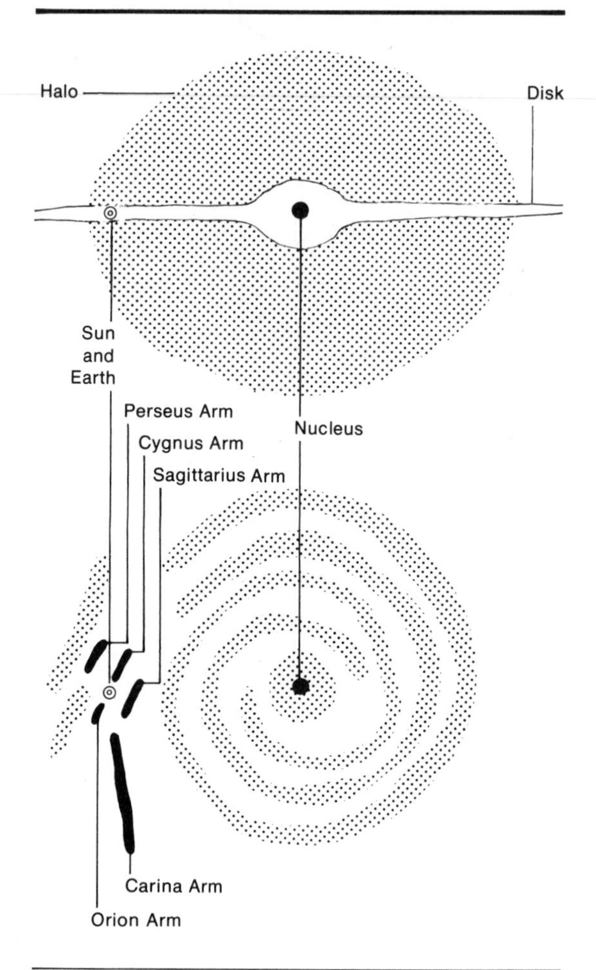

FIGURE 7-2
Two views of the Milky Way galaxy. The shaded spiral arms are drawn in schematically; their exact positions are not known.

schematic view, one way of putting together a spiral pattern out of the radio astronomers' observations.

The center of the disk contains a central bulge, within which is the galactic core. The core is quite a dramatic place, because the stars are very close together. If you lived on a planet circling one of the stars in the core, you would see a nighttime sky that would be almost as bright as the twilight sky here on earth. Some of the stars would be near enough to show a disk, perhaps. We are just beginning to unravel some of the mysteries of the galactic core. (This part of our galaxy will be discussed further in Chapter 10.)

The galactic disk, containing the spiral arms and the nucleus, is embedded in the halo. This halo contains old stars, many contained in globular clusters of stars, in contrast to the younger stars in the spiral

FIGURE 7-3 The spiral galaxy Messier 51. (Hale Observatories.)

arms in the disk. How can you tell which stars are old? Massive, hot, young blue stars burn their nuclear fuel at a prodigious rate. As their fuel is limited, they do not last very long. After a few million years, a short time compared with the ten-to-twenty-billion-year age of the Galaxy, these blue stars burn out, ending their stellar lives. Thus a region of the galaxy containing blue stars also contains young stars, recently formed. The spiral arms are such a place, where star formation is still going on. The halo, in contrast, contains no young, blue stars, just old, red stars, which have been able to last until now because they do not use up their fuel very fast.

The core, disk, and halo represent three different parts of our galaxy. Such a division is also reflected in the shapes of other galaxies.

Galactic forms

Our galaxy is one of billions of galaxies in the universe. Only three of these billions of galaxies are visible to the naked eye – the two clouds of Magellan visible only to residents of the Southern Hemisphere and the Andromeda galaxy, a faint fuzzy patch in the autumn sky. All the other galaxies are telescopic objects. Some were discovered by Charles Messier, an eighteenth-century astronomer who compiled a list of 103 nonstellar patches of light to help him search for comets; the Andromeda galaxy is thus known as Messier 31, or M 31. Fainter galaxies were discovered somewhat later.

Anyone who seeks to understand galaxies must first classify them, and the most obvious way to classify galaxies is by their shape. It is assumed that galaxies with similar shapes are similar in other respects also: size, total brightness, stellar content, and evolutionary history. The three principal shape classifications are spiral, elliptical, and irregular.

Spiral galaxies have much the same shape as our Milky Way. Messier 51, shown in Figure 7-3, is a somewhat more loosely wound spiral than ours, but the overall appearance is similar. The spiral arms of Messier 51 contain a lot of gas, dust, and young stars, as the arms of our galaxy do. Color photographs show that the spiral arms are somewhat blue, as you would expect if they contain young stars. The blob at the end of one of the arms of M 51 is a companion galaxy, and current theory indicates that the gravitational interaction between this companion and the main galaxy is at least partly responsible for the existence of the spiral pattern. Some spiral galaxies have bars in the middle, with the spiral arms attached to the ends of the bars, while others have rings at the centers of the arms. We are a long way from understanding how these structural features of spiral galaxies originate.

If you consider a spiral galaxy without its halo and nucleus, only the spiral arms remaining, you have a good idea of the stellar content of an irregular galaxy. The Magellanic Clouds, landmarks of the southern sky, are both irregular galaxies. Figure 7-4 shows both of them. The Large Magellanic Cloud has a trace of a one-armed spiral structure, which you can detect in the photograph if you look at it for a long time. These clouds are composed primarily of blue stars, and contain a great deal of gas and dust. They are both small, much smaller than our own galaxy; this is true of most irregular galaxies.

Now consider a spiral galaxy without the disk and spiral arms.

FIGURE 7-4 The Magellanic Clouds, with the Large Cloud at the left. This is a wide-field view of the sky, as the Large Cloud is twelve degrees across. The bright star at the upper right is Achernar, a star in our own galaxy. (Harvard College Observatory.)

The halo and nucleus remain, and you have an approximate idea of what the third main type of galaxy, the elliptical, looks like. An elliptical galaxy resembles the halo and nucleus of a spiral galaxy in that it is roughly spherical and has quite old stars in it. Elliptical galaxies present a variety of shapes, from spherical galaxies that look circular from our vantage point to flattened ones. A photograph of a dwarf elliptical galaxy, one of the flattened ones, is shown in Figure 7-5. It is evident that there is no dust, and radio observations show very little gas.

Two basic types of elliptical galaxies exist, the dwarfs and the

FIGURE 7-5 NGC 205, a dwarf elliptical galaxy. Note that individual stars are visible. (Hale Observatories.)

giants. Dwarf elliptical galaxies are only a few kiloparsecs across, much smaller than the 30-kiloparsec diameter of our galaxy; NGC 205, the dwarf elliptical shown in Figure 7-5, is 4.2 kiloparsecs across. Giant elliptical galaxies, about 50 kiloparsecs across, also exist, and many of these ellipticals are similar in properties to the quasars.

MASSES OF GALAXIES

So far, the sizes of galaxies have been discussed as "linear" sizes, how much space the galaxy occupies. The mass of a galaxy is also an important quantity to know, since the mass of a galaxy has some effect on its evolution and its structure. Masses of galaxies are not as well known as some of the other properties of galaxies; galactic masses cannot be directly measured.

A variety of observations of our own galaxy, both direct and indirect, indicate that our sun is revolving around the galactic center, traveling at a speed of 250 kilometers per second. The galaxy is so large that it takes us 2.4×10^8 years to complete one trip around it, and we have completed only 20 trips since the sun was formed. We are held in our orbit around the galactic center by the gravitational force of the mass in that part of the galaxy inside our orbit. We can play the same game that we have played before in connection with binary stars; the gravitational force acting on us must be enough to keep us in orbit, and not too much lest we fall straight towards the galactic center. A straightforward calculation indicates that the mass of our galaxy is about 2×10^{11} solar masses. To calculate more accurately, one must consider the mass outside the sun's orbit and in the galactic disk, and the effects of the distribution of mass in the galaxy.

We can measure the masses of other spiral galaxies by much the same method. A spectrum of a spiral galaxy can be analyzed and the Doppler shift of the spectrum lines at different points in the galaxy will give the rotational velocity of the galaxy at different points in it. Analysis of this rotation curve then will give the mass of the galaxy. A primary shortcoming of this procedure is that the outer parts of the galaxy, which are too faint to allow the spectrum to be obtained, are neglected in the analysis. A number of spiral galaxies have been analyzed by this rotation-curve method, and their masses vary from 2×10^{10} to 4×10^{11} solar masses. Apparently the Milky Way, with its mass of 2×10^{11} suns, is a fairly normal spiral galaxy.

Elliptical galaxies do not rotate, so the rotation-curve method cannot be used to determine their masses. There are many galaxies that are apparently doubles, and measurements of the orbital velocity of these galaxies around each other allow measuring the masses of some ellipticals. One measures the relative velocity of the two galaxies in a pair, and the necessary balance between gravitational force and orbital motion allows one to determine a mass. Values for the masses of spiral galaxies measured by this method roughly agree with the values obtained by the rotation-curve method, considering the large uncertainties involved. The double-galaxy method does allow the masses of giant elliptical galaxies to be measured, and results indicate that the giant ellipticals are probably the largest galaxies known, with masses of roughly 10^{12} solar masses.

The double-galaxy method and the rotation-curve method for measuring masses of galaxies have one shortcoming that two Princeton investigators, Jeremiah Ostriker and P. James E. Peebles, recently pointed out. Both of these methods use the motion of one object around another to determine a mass. Any spherical distribution of mass, or massive halo, around the objects and outside of the zone of their motion will not affect the rate at which the orbits are completed. For example, our galaxy could be surrounded by a massive halo and we should never detect the existence of the halo by analyzing motions in the galactic disk. While the existence of these halos is conjectural at present, we may be making a serious error in our mass estimate by neglecting them.

But, you ask, shouldn't we see this halo? No, we shouldn't see it if very small, low-mass stars composed it. Such stars are very dim and hard to see. Specifically, if the halo were made of 0.1-solar-mass stars, which have 2×10^{-3} times the luminosity of the sun, it would be one-fiftieth as bright as it would be if it were made up of one-solar-mass stars. Stars of 0.1 solar mass have high mass-to-light ratios (defined as the star's mass divided by the star's luminosity in units where the sun equals 1, so a 0.1-solar-mass star has a mass-to-light ratio of 50). You can hide a great deal of mass in small stars with high mass-to-light ratios, since such stars are visible only if they are close to the sun.

We shall return to some of these uncertainties in measuring the masses of galaxies in Chapter 16, where it becomes important that we know the masses of galaxies fairly precisely. For present purposes, it is only important that you have a general idea of the properties of galaxies as a background to the quasar story.

SUMMARY OF GALACTIC PROPERTIES

Figure 7-6 summarizes the properties of galaxies. The numbers given there are intended to represent typical values and not values for

DIAMETER (KILOPARSECS)	30	6	3	50
MASS (SOLAR MASSES)	10^{11}	10^{9}	Poorly Known	10^{12}
LUMINOSITY (SUNS)	10^{10}	10^{9}	10^{8}	10^{10} to 10^{11}
COLOR	Blue (Disk) Red (Halo and Nucleus)	Very Blue	Red	Red
GAS CONTENT	5%	15%	<1%	<1%
TYPES OF STARS	Young (Disk) Old (Halo)	Young	Old	Old
EXAMPLE	M 51 (Figure 7-3)	Magellanic Clouds (Figure 7-4)	NGC 205 (Figure 7-5)	M 87 (Figure 10-3)

FIGURE 7-6 Properties of galaxies.

any individual galaxy; they are approximate. To understand quasars, you need only to understand what galaxies are like in a very general way.

Groups of galaxies

Galaxies are found everywhere in the universe. The only place where it is hard to find them is in the direction of the Milky Way, because the stars and dust in our own galaxy obscure the view. Galaxies are scattered throughout the universe, each one an island of many, many stars. How do we know that they are so far away?

It took the power of the Mount Wilson 100-inch telescope to show us individual stars in galaxies. Look closely at Figure 7-5 and you will see that you can detect the images of individual stars in this galaxy, one of the companions to the Andromeda galaxy. If you can recognize some of these stars as familiar stars, you can then compare them for brightness with similar stars in our own galaxy. Such a comparison gives a distance of 700 kiloparsecs to NGC 205, roughly the same as the distance to the Andromeda galaxy. Light takes 2.2 million years to travel from these galaxies to us. Yet NGC 205 is one of the nearest galaxies.

The 18 nearest galaxies form a loosely organized collection known as the Local Group. The composition of the Local Group should give you some idea of the types of galaxies that exist in the universe. The two largest galaxies are the Milky Way galaxy and Messier 31. There are two other spirals, M 33 and Maffei 1, which was discovered just recently since it is located in the direction of the Milky Way, hidden by dark dust lanes. The lesser members of the Local Group are the most numerous: four irregulars and ten dwarf ellipticals.

At least half of all galaxies, and perhaps all, are found in clusters of some sort. Some clusters of galaxies are small, loose collections like our own Local Group, while others are considerably larger. The nearby groups are small groups like the Local Group, so it is these groups that we know most about. One needs to go a bit farther away to find the larger clusters, which contain the giant elliptical galaxies. The nearest large cluster is the Virgo cluster, named for the constellation that it lies behind. This name may be misleading, for the stars that make up the constellation Virgo are much, much closer than the distant cluster of galaxies. The Virgo cluster's brightest member is one of the nearest giant elliptical galaxies, Messier 87. This Cadillac of galaxies is a hundred times as massive as our galaxy and is a source of intense radio-frequency radiation. The Virgo cluster of galaxies, one of the largest, most populated clusters, contains hundreds of different galaxies. The part of the sky containing the Virgo cluster is known as "the realm of the galaxies" to amateur astronomers. I am told that it is an inspiring sight if you are lucky enough to have a 12-inch telescope and a dark sky.

The expanding universe

As we probe further and further into the depths of intergalactic space, the spectra of more and more distant galaxies are increasingly redshifted, indicating that these galaxies are moving away from us. The speed of recession increases with the distance of the galaxy from us. Figure 7-7 shows a series of photographs of galaxies and their spectra. As your eye travels down the page, you note increasingly distant galaxies; their apparent angular size and their apparent brightness both decrease

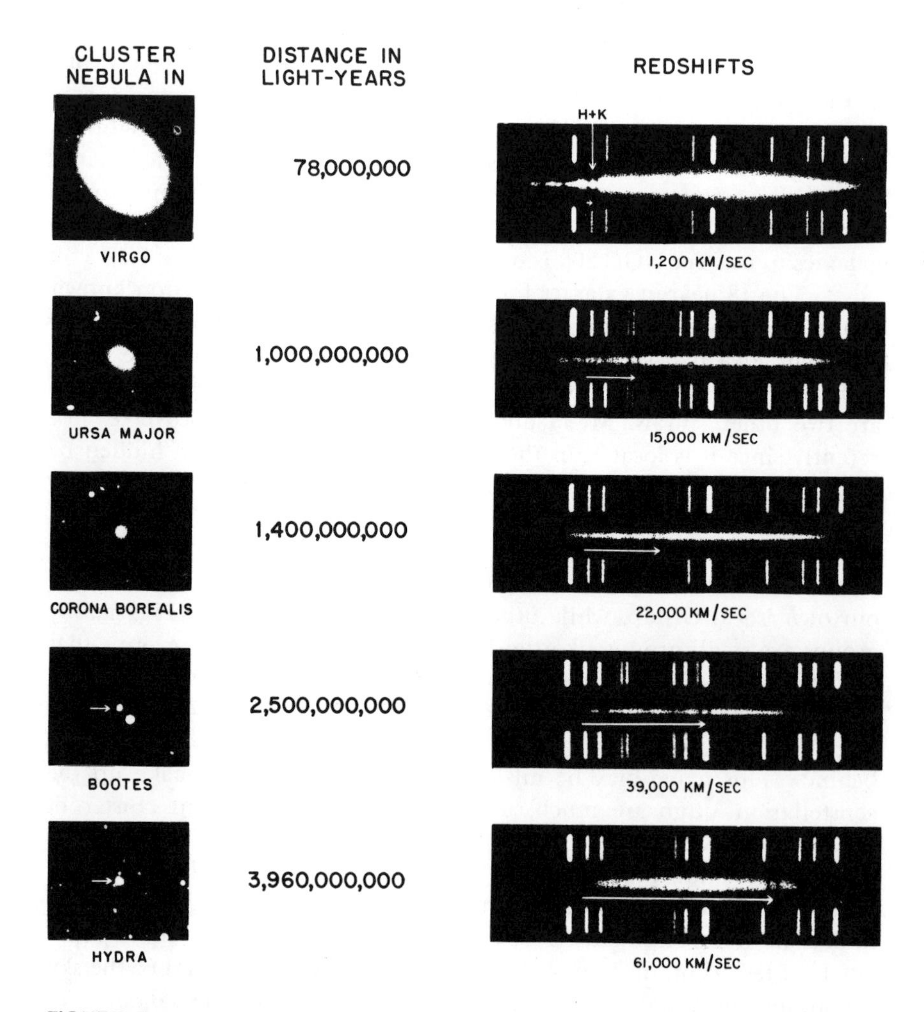

FIGURE 7-7 An illustration of Hubble's Law: The redshift increases with increasing distance. (Hale Observatories.)

as their distance from our galaxy increases. In itself, this is not surprising; it is quite natural. But look at the spectra. The red wavelengths are on the right-hand side of the page, the blue wavelengths on the left. The two dark lines visible in each spectrum are characteristic of galactic spectra, the so-called H and K lines of ionized calcium. They are found at the extreme blue end of the spectrum unless the redshift is large. This is the case with the galaxy at the top of the page, in the Virgo cluster, where the redshift is quite small. Farther and farther out in the universe, the redshift becomes larger and larger.

The galaxies only seem to be moving away from our particular galaxy; in fact, the galaxies are all moving away from each other as a result of the expansion of the universe. The velocity of recession and the distance of a galaxy are correlated: the greater the velocity, the larger the distance. This relation was discovered by Hubble and Humason in the 1930's, and it is known as Hubble's Law. This relation can be written: Velocity of recession equals Hubble's constant times distance. Thus, to measure the distance to an extragalactic object, you measure the redshift, and Hubble's Law gives you the distance. Such a procedure works as long as the redshift is due to the expansion of the universe. Hubble's constant has been measured many times with differing results; the currently accepted value is 50 kilometers per second per megaparsec. The measurements on which this number is based will be discussed in more detail in Chapter 14; the exact value of the Hubble constant is not a prerequisite to understanding quasars. The increase of redshift with distance has been well established for ordinary galaxies, but the objects with the largest redshift, and therefore presumably the greatest distance, are the enigmatic quasars.

Quasars

A quasar is by definition a starlike object with a large redshift. Quasars were originally discovered as a source of radio emission. The name originated from the letters QSRS, standing for "quasistellar radio source." Several objects that look similar (although they are not sources of radio emission) are known to exist, and the name *quasar* has been applied to all of them. The term is used to describe all objects that look like stars on photographs taken with the 48-inch Schmidt telescope at Palomar and that have spectra that show large redshifts. (The spectra of ordinary stars show very small redshifts.) The term *quasar* has replaced the confusing variety of terms used in the literature of the 1960s: QSO (quasistellar object), QSS (quasistellar source), BSO (blue stellar object), interlopers, and QSRS.

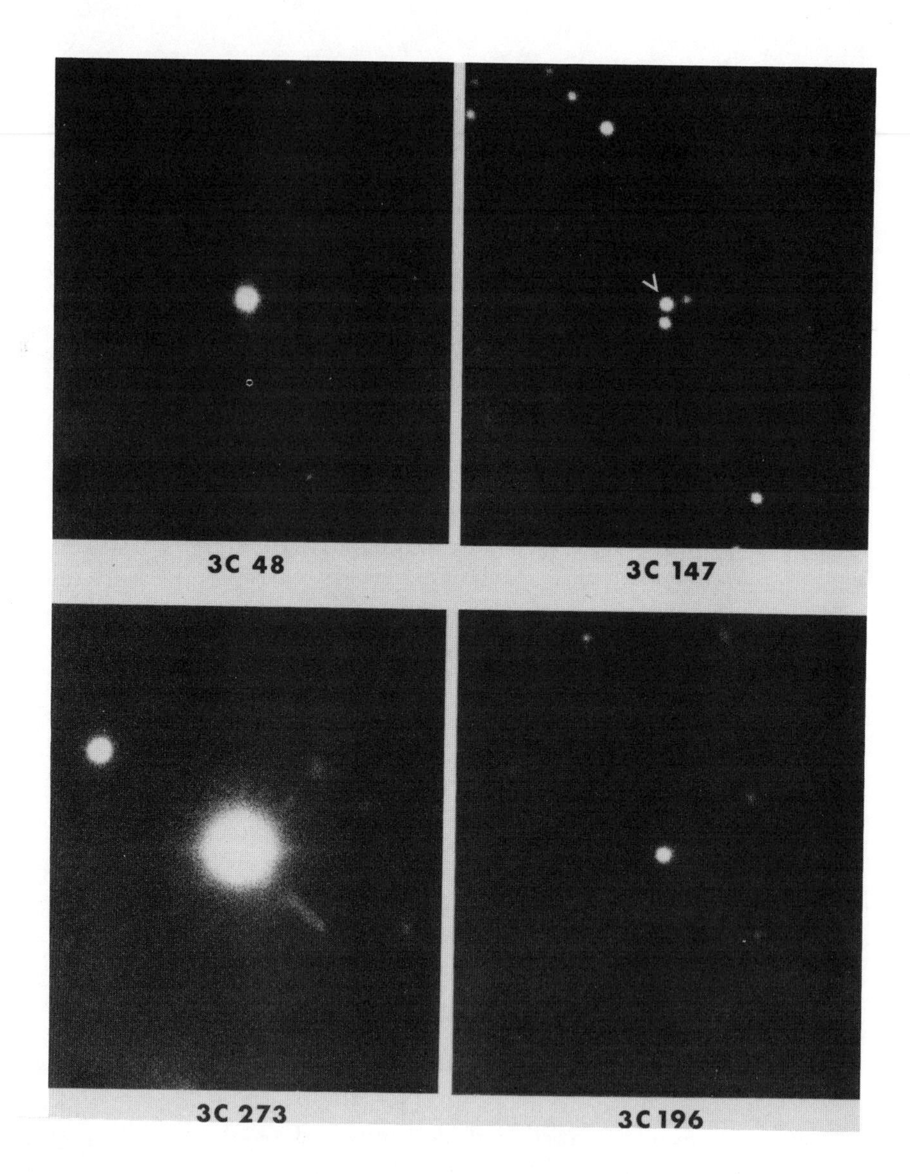

FIGURE 7-8 Four quasars. (Hale Observatories.)

A photograph of four quasars is shown in Figure 7-8. The object 3C 273 is the brightest quasar known. Although 3C 273 is a twelfth-magnitude object, visible in smallish telescopes, its peculiar nature was undiscovered for a long time since it looks like a star. There are about ten million stars of comparable brightness in the sky, and it is not surprising that this remarkable object remained unnoticed for a long time.

One of the other quasars in Figure 7-8, numbered 3C 48, also was important in the discovery of these objects.

IDENTIFICATION OF QUASARS

In the 1950s, radio astronomers discovered a multitude of objects in the sky that emitted large quantities of radio-frequency radiation. Because radio telescopes cannot easily pinpoint the location of a radio source (for reasons described in Chapter 8), it was not known what sort of visible object, if any, corresponded to the radio sources. A few radio sources allow their positions to be determined more accurately as the moon passes in front of them and cuts off the radio waves. If you know the moon's orbit, you can measure exactly when the moon cuts off the radio noise and get a much better idea of where the radio source is in the sky. Identification of radio sources is also easier if the source lies far from the Milky Way, for then there are fewer stars near the position of the radio source.

The first quasar to be recognized as a strange object was 3C 48. Thomas Matthews and Allan Sandage of the Hale Observatories noticed an unusual stellar object at the location of this radio source, number 48 in the 3C catalogue prepared by the Cambridge University radio astronomers. Several astronomers obtained spectra of this object and saw something very strange: broad emission lines at wavelengths that did not correspond with any features normally seen in spectra of other objects. Such a spectrum is not at all like the spectrum of a star. These emission lines meant that this object was concentrating its photon emission at certain wavelengths, but these were not the wavelengths normally seen in spectra of other objects in the galaxy.

The solution to the 3C 48 puzzle came a few years later. In 1962, the moon passed in front of another radio source, 3C 273. Cyril Hazard and his Australian colleagues were able to time this occultation and to determine accurately the position of the radio source. It was then known which one of the many stars near the radio-source position was producing the radio noise, and it was a bright one, the brightest quasar ever discovered. California astronomer Maarten Schmidt then obtained a spectrum of this quasar, and noticed the same peculiarities that were noticed with 3C 48. Once again, 3C 273 had a bright-line spectrum, quite unlike the usual stellar spectrum. It was emitting light at only a few wavelengths, and these wavelengths fitted no pattern. Hydrogen, the most common element in the universe, will emit light at just a few wavelengths when it is observed in a glowing gas cloud, but such were not the wavelengths seen in 3C 273.

Schmidt puzzled over his spectrum for a while. As he was writing a report on the results, on February 5, 1963, the answer appeared in one of those flashes of intuition that are the supreme rewards of a successful scientific career. He noticed that the emission spectrum of 3C 273 did correspond to the hydrogen spectrum if he assumed that this starlike object had an enormous redshift — a redshift corresponding to a recession velocity of 47,000 kilometers per second, more than one-tenth the speed of light. No star in the galaxy could possibly move that fast; it would have escaped from the galaxy long ago. Furthermore, a star would not have an emission spectrum like the one of this object. By 1974, the spectra of over two hundred quasars had been analyzed, all of them having very large redshifts. Such redshifts characterize the quasars and give them their mysterious nature.

The redshift of a quasar is usually denoted by the letter z, which equals the shift in wavelength of a spectrum line ($\Delta\lambda$) divided by the wavelength that that line had when it left the quasar (λ_0; thus $z = \Delta\lambda/\lambda_0$). It can also be expressed as a velocity by the Doppler shift formula. Quasar redshifts range from relatively small numbers, like 0.15 for 3C 273, to large, like 3.53 for OQ 172, the most distant quasar known at this time (OQ = Ohio State University radio survey, List Q). The simplest way to explain these redshifts is to assume that the quasars are extremely distant objects that follow Hubble's Law, and are thus the most distant objects known. Most astronomers accept this explanation, and the bulk of Part Two will be based on it. Alternative explanations for the redshift will be considered in Chapter 12.

PROPERTIES OF QUASARS

If the redshifts of quasars are caused by the expansion of the universe, they are very luminous objects indeed. The most powerful one, 3C 273, is five trillion times as luminous as the sun. Other quasars are comparable (Table 7-1). You can see 3C 273 in a four-inch telescope, even though it is 900 megaparsecs away. The light you see from 3C 273 left the quasar almost three billion years ago, when no land life existed on the earth. It would take a 60-inch telescope to spot a giant elliptical galaxy, were the elliptical as far away as 3C 273 is.

As the name indicates, many quasars are radio sources. The total amount of energy emitted in the radio range is somewhat less than the optical luminosity, and in some cases the radio emission is much less, being undetectable. Recent radio measurements indicate that the radio emission comes from very small, compact clouds of high-speed electrons producing synchrotron radiation as they gyrate around a magnetic field.

The radiation from quasars in the optically visible part of the electromagnetic spectrum is not all in the form of emission lines from a cloud of hot gas. Some of this radiation is not concentrated at any particular wavelengths. Such continuous radiation is also seen in the infrared

TABLE 7-1 Properties of Quasars

QUASAR	REDSHIFT, z	VISUAL MAGNITUDE	RADIO BRIGHTNESS *	DISTANCE (MEGAPARSECS)	VARIABILITY	LUMINOSITY (SUNS)
3C 48	0.367	16.2	47	1700	Occasionally	2×10^{12}
3C 147	.545	16.9	58	2600	None detected	2×10^{12}
3C 273	.158	12.8	67	900	Yes, on time scales of years	5×10^{12}
3C **196**	.871	17.6	59	3200	Yes	1×10^{12}

* In flux units; 1 flux unit equals a brightness of 10^{-26} watts per square meter per hertz at the surface of the earth.

—it may be that most quasars put out most of their radiation in the infrared part of the spectrum.

The continuous radiation from quasars is variable with time. Most of the observations of variation have been made in the optical part of the spectrum. Most quasars vary relatively slowly, brightening or dimming over periods of a year or so, but a few are considerably more violent in their variations, doubling their brightness in periods of a day or so.

Each of these properties is worth considering in turn, as we try to find how studies of different types of radiation from quasars can indicate their nature. Most people believe that quasars are related to galaxies in some way. But exactly what causes a galaxy to become a quasar is not known. There are some theoretical speculations about the nature and evolution of quasars (Chapter 11), but most of our knowledge about quasars comes from explanations of the observations and not from any detailed theoretical scheme. The study of quasars is still basically an observational science.

Most of the objects in the extragalactic universe are galaxies like our own. Every galaxy is a collection of myriads of stars, held together by gravity. They come in various sizes and shapes, from the small dwarf elliptical galaxies to the giant elliptical galaxies, ten times larger than our own galaxy, the Milky Way.

Yet galaxies are not the most luminous objects in the universe. That honor belongs to the quasars, which are hundreds of times as luminous as galaxies. This chapter just touched on the properties of these strange objects, and the chapters that follow will describe them in more detail.

RADIO WAVES FROM QUASARS

Quasars were originally discovered because they were intense sources of radio radiation. Any complete model of a quasar must be able to explain where this radiation, which exceeds the optical luminosity of our galaxy in its power, comes from. Recent observations with pairs of radio telescopes on opposite sides of the world indicate that the source of this radio noise is some object a few light-months across. How can such a small volume produce so much energy?

Before considering detailed models for the radio-emitting region, we must first ask how we know that the region is so small. Here we encounter the vast difference between the techniques of optical astronomy and radio astronomy. One difference, relevant to this matter, is that a single radio telescope cannot focus radio waves as well as an optical telescope can focus light. As a result, a single radio telescope gives a very fuzzy view of the radio sky, but pairs of radio telescopes can be combined to sharpen the perception.

In addition, the source of radio waves is often very different from the source of light waves. Light from stars is generated by the heat of the gas that makes up a star. Such thermal radiation can, in some cases, produce radio noise, but the radio waves from quasars come from a quite different source — electrons spiraling around magnetic lines of force at speeds close to the speed of light. Two principal questions are associated with the radio emission from quasars: How do we know that the radio-emitting region is so small? What sort of mechanism is responsible for the radio noise?

Fuzzy views of the radio sky

Radio astronomers face the disadvantage that their telescopes give a very fuzzy picture of the radio sky — somewhat like the view of the world that a nearsighted person gets when he takes his glasses off. You can measure the sharpness of any picture of the sky by noting its resolution, which is the size of the fuzzy, blurred image of a pointlike object, or equivalently, the distance by which two objects need to be separated to be perceived as separate objects. It is the limited resolution that prevents a person with eyeglasses removed from distinguishing the letters on the bottom lines of the eye chart. They look like blurs.

The resolving power of a telescope, or the human eye, is measured in angles, as size for all images is expressed in angles. The size of the moon in the sky is half a degree, or 30 arc-minutes. A single radio telescope generally has a resolving power of about 30 minutes of arc, so that a radio view of the sky, from a single telescope, would show many blobs the size of the full moon. The human eye can do much better than this, as its resolution is about one minute of arc. If you want to see what a fuzzy view of the sky looks like, you can get it if you are nearsighted. Just take off your glasses and go outside and look up. You will be amazed at what you cannot see. Such is the fate of the radio astronomer who has not discovered interferometry, or the art of combining pictures from two telescopes to obtain a sharp representation.

Radio telescopes have poor resolving power because of the long wavelength of radio waves. Light waves are very short, with wavelengths of less than one micrometer (1 micrometer $= 10^{-6}$ meter or 1/25,000 of an inch). Radio waves have wavelengths ranging from a few centimeters to some tens of meters. A telescope can focus well only if its dimensions are considerably greater than the length of the waves that it is focusing. This condition is satisfied with optical telescopes, which are several meters across, large compared with the small length of light waves, but it is difficult to satisfy this condition with a radio telescope unless it is thousands of meters — kilometers — in diameter. Building a single radio telescope that was a few miles in diameter would be a monumental (and expensive) undertaking.

To understand why telescopes have to be so much larger than the wavelengths that they are focusing, consider a somewhat simpler problem. Instead of worrying about focusing, just worry about how sharp a shadow an object can cast. If something can cast a sharp shadow, it can also focus well. Light shadows are quite sharp, because walls and trees and buildings and people, objects that generally cast shadows, are much bigger than light waves. Such light waves can also be well focused by people-sized objects, like telescope mirrors. But think of the shadows that breakwaters cast, as they try to cast "shadows" in their preventing ocean waves from reaching into a harbor. These shadows are not sharp, but fuzzy. Immediately behind the breakwater, the water is quite calm (Figure 8-1). However, soon the water waves manage to turn the corner, around the breakwater, and the area of calm water has a fuzzy boundary. Sound waves do the same thing, for they can turn corners around buildings. The reason that sound waves and water waves can turn corners around obstructions and cast fuzzy shadows is that their wavelengths are

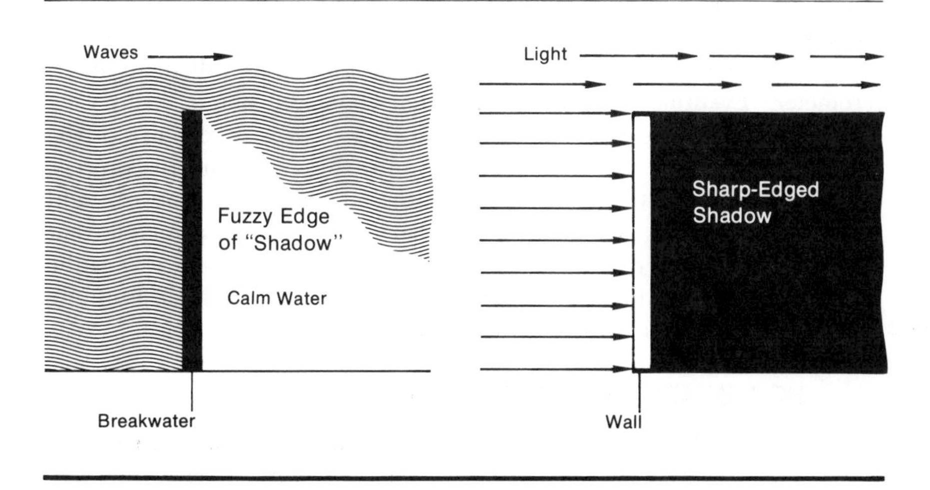

FIGURE 8-1 As water waves approach a breakwater from the left, the breakwater is unable to cast a sharp shadow since the waves eventually turn the corner and penetrate the space behind the breakwater.

long – a few feet in both cases. But light waves are short, and cast very sharp shadows. The upshot is that light waves, with their short wavelength, are easy to focus, while long-wavelength radio waves are not.

For a first look at the radio sky, fuzzy pictures were not bad at all. Radio astronomy pioneers wanted to find out whether there was anything to be seen at all before trying to get sharp pictures. The first maps of the radio sky, obtained by Karl Jansky of Bell Laboratories, the man who discovered that radio emission from the sky existed, and by the American amateur Grote Reber, were quite fuzzy, only good enough to indicate the presence of a few blurred concentrations of radio emission. Some of these concentrations came from obvious places, like the galactic center. But other sources of radio noise had no obvious cause. You cannot go to an optical astronomer and say, "There's a radio source in the constellation Cygnus – what do you think it is?" You need better positions for radio sources, and to get them you need a sharp picture.

Interferometry

You could, of course, use the brute force method and build a radiotelescope a few kilometers in diameter if you could persuade someone to foot the billion-dollar bill. Yet there are ways to be a little more clever than that. You can use a pair of radio telescopes a few miles apart to obtain some of the information that you could get from a huge dish. Because you combine the signals from the two dishes and watch the signals interfere with each other, such a combination is called an interferometer. Eventually, a pair of radio telescopes can obtain a sky view equivalent to the view obtained by a single telescope of the size of the distance between the two telescopes. If you wish to speed up the process, so you do not have to spend three months on one object, you can add more radiotelescopes and create a number of pairs. The principle that governs the working of all interferometers or multitelescope arrays is the same, and is easiest to understand in the context of the basic building block of an array — a two-element interferometer.

HOW INTERFEROMETRY WORKS

Figure 8-2 shows a two-element interferometer, observing radio waves with a wavelength of ten centimeters and with a separating distance of three kilometers between the two dishes. Both antennas are looking at the radio source simultaneously, as the source passes overhead. The signals from the source come in waves, which you can imagine as consisting of crests and troughs, alternating. (For electromagnetic fields, these are not really crests and troughs like water waves, but the analogy is still reasonable.) The signal picked up by each antenna is displayed in the figure and then fed into a mixer, which combines the two signals and displays the output. The mixer watches the two waves interfere with each other; hence the name interferometer.

This antenna seeks to determine the exact location of a point source of radio waves by determining exactly when the source is overhead. At that time, in Figure 8-2 (a), crests in radio waves from the source reach both antennas at the same time. These signals are in phase, and when the two separate signals are added together, the signal strength is increased. An ordinary radio telescope would continue to see a strong signal for a long time, since it would see the source as a large blob. But our pair of telescopes can pinpoint the location of the source much more accurately. As the source moves across the sky, it soon reaches a point (Figure 8-2 (b)) where a trough from the left-hand telescope reaches the mixer at the same time as a crest from the right-hand telescope. The two signals mix and cancel, and zero signal comes out of the mixer. We now know that the source is no longer overhead. It seems now that we can pinpoint the source, for a pair of telescopes three kilometers apart can locate the source within a few seconds of arc.

One measurement will not quite do the job, however. Look at

FIGURE 8-2 How an interferometer works. (See text for description.) The horizontal scale has been compressed for the sake of clarity.

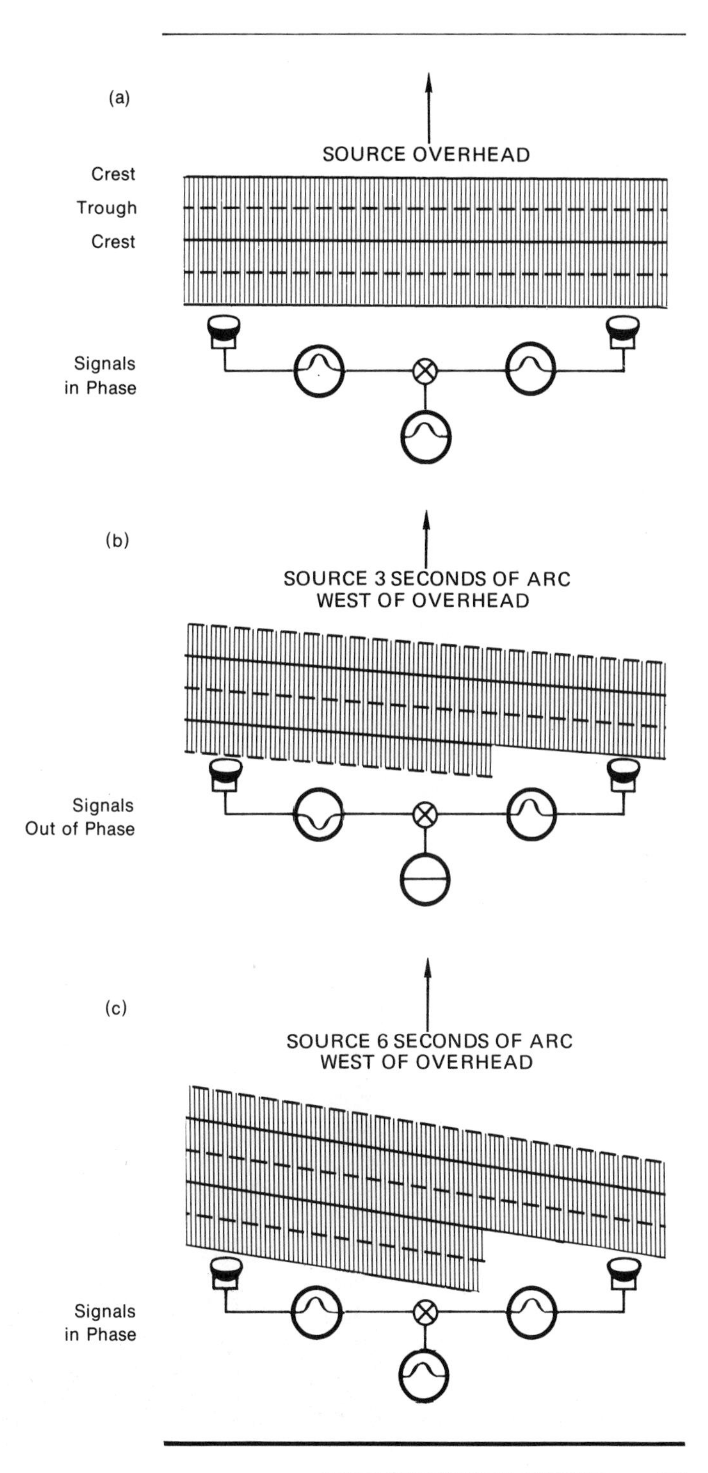
(a)
SOURCE OVERHEAD
Crest
Trough
Crest
Signals
in Phase
(b)
SOURCE 3 SECONDS OF ARC
WEST OF OVERHEAD
Signals
Out of Phase
(c)
SOURCE 6 SECONDS OF ARC
WEST OF OVERHEAD
Signals
in Phase

Figure 8-2 (c). The source has moved a little farther along, and once again two crests reach the mixer simultaneously. The signals add again. It is not the same crest that reaches the mixer; however, someone who can see only the output from the mixer cannot tell whether it is the same crest or two different crests that add. How is the observer to tell that the source is not overhead now? If he were to draw a map of this source, he could say only that the source either is overhead now or was overhead a short while preceding, when the same crest arrived at both scopes. Such a map would look like Figure 8-3. The white stripes indicate the parts of the sky that contribute some signal to the interferometer when the antenna looks at them. If the source is a single point, all that can be said is that it lies in one of the light stripes. Which stripe is still doubtful. The reason that the stripes are elongated in the north-south direction is that our simple interferometer is no better than an ordinary radio telescope in a direction perpendicular to the line between the two telescopes, and it still sees a blur.

What the interferometer has done is to take the blurred picture of a radio source provided by a single telescope and slice it into a series of stripes, as in Figure 8-4. It is somewhat difficult at this point to imagine what you might do with such a striped picture of the radio sky. It doesn't seem to be much better than a blur. This striped pattern is known as the interferometer's beam pattern. The interferometer has told us that the radio source we are seeking is located somewhere in one (or more) of those stripes.

The usefulness of these striped pictures of the radio sky can be appreciated a little better if you go back to our basic challenge. We know that there is a radio source up there somewhere and we are trying to find

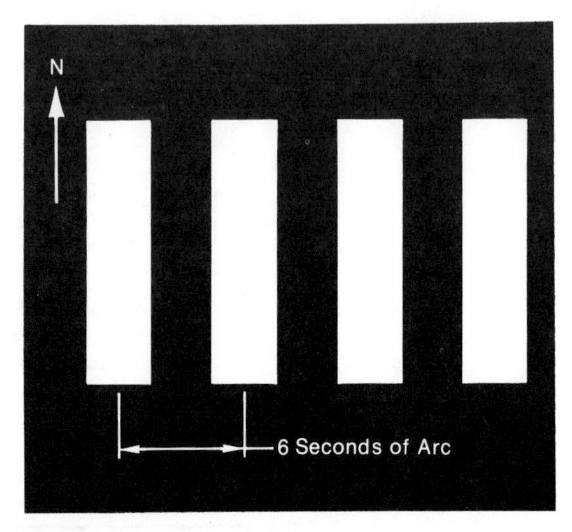

FIGURE 8-3
A map of the radio source detected by the interferometer of Figure 8-2.

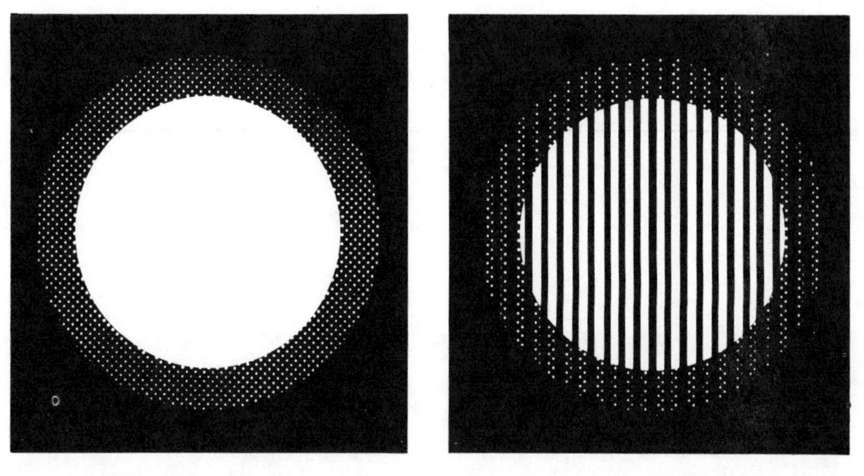

FIGURE 8-4 A comparison of the beam patterns of an interferometer and a single radio telescope.

out where it is. With a single telescope all we can do is take a sky chart, draw a big circle on it, and tell our optical astronomer friends that the source is somewhere in that circle. They will then tell us to go home and do better, as there are thousands of stars in that circle and they certainly aren't going to look at them all to see if they can find the optical counterpart to the radio source. With an interferometer, we can draw some long skinny boxes on the map and say that the source is in one of those boxes. If we can now overlay our first set of boxes with a second set of different boxes, maybe we can get somewhere. Consider a small part of the beam pattern to see how this procedure works.

Figure 8-5 demonstrates this procedure. The left panel shows a sky map of the part of the sky in which the radio source lies. It shows four stars (A, B, C, and D) and four galaxies, (1, 2, 3, and 4). We want to know which one, if any, is the radio source. A single radio telescope would tell us just that the radio source lies somewhere on that map, which isn't much help. One interferometer would slice the map into stripes, as shown in the second panel of the figure. It provides some help, as we can rule out galaxies 2 and 4 and star B as a possible source. But this is not good enough. We then use a second interferometer, whose telescopes are at right angles to the first. We could also adjust the spacing between the telescopes to obtain wider stripes than the first interferometer had. This second interferometer rules out galaxy 1 and stars A, B, C, and D. Putting the two pictures together, we see that if

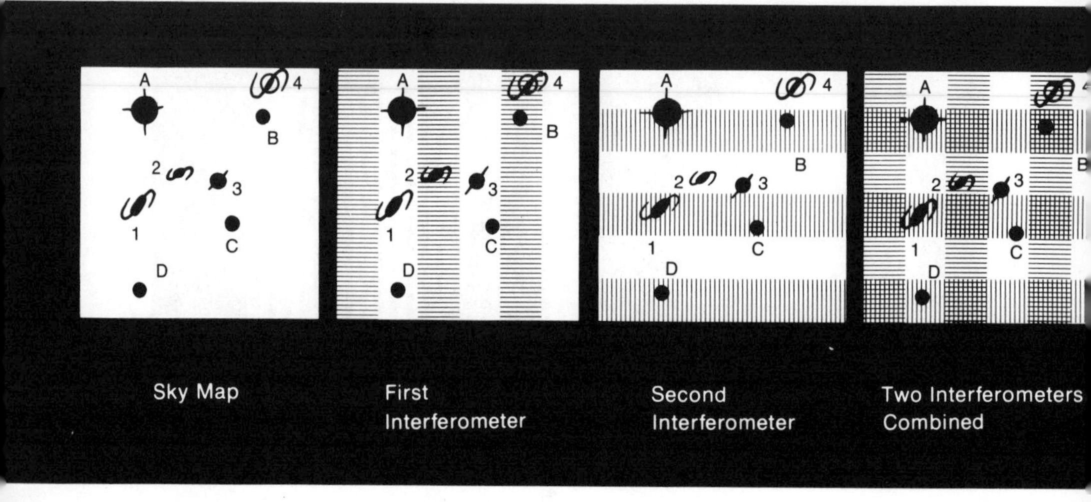

FIGURE 8-5 How two interferometers combine to provide a precise location for a radio source. (See text for details of the procedure.)

the radio noise is coming from any of the objects on the star map, it must be coming from galaxy 3 (right-hand panel). We can now go and tell the optical astronomers that galaxy 3 is a good candidate for the optical counterpart to the radio source.

If we wanted to make sure that galaxy 3 was really a radio galaxy, and that the emission was not coming from some invisible object in one of the white areas on the right-hand panel of Figure 8-5, we should have to go back and take some more readings with different spacings. When you change either the length or the orientation of the interferometer's baseline, the line connecting the two antennas, you change the width or orientation of the stripes in the beam pattern. If an interferometer is to provide a complete map of an area, many different antenna combinations must be used, and all of this switching around takes much time, even though the rotation of the earth automatically changes the stripe pattern as the source moves across the sky.

It seems like a laborious way to go about finding radio sources. The keys to success are your ability to use radio telescopes in tandem rather than individually, and your ability to move the telescopes around to create many, different striped pictures that can be combined to pinpoint the radio source.

You are probably asking whether there is an easier way to do it. If you have an array of radio telescopes, each one can be combined with each of the others as one individual interferometer. The 1970s may see the construction of such an array on the Plains of Saint Augustin in New Mexico (see Figure 8-6). The plans call for an array of 27 different telescopes spread in a carefully calculated pattern 26 miles in diameter. Each telescope can combine with each of the others to produce a single striped picture, and the central computer can combine all of these pictures to produce a good sharp picture in a day or so, compared with the month-long procedures of two- or three-telescope interferometry. We do not yet have the Very Large Array, VLA, as it is called. Construction has begun and the instrument has been funded, and in the absence of any unforeseen disasters it will be fully operational by 1981.

TRANSCONTINENTAL RADIO ASTRONOMY

Although the VLA would be an excellent instrument for providing high-quality maps and positions for large numbers of radio sources, we can, in a sense, obtain even more sharply focused pictures from larger interferometers, which are made up of radio telescopes thousands of miles apart. The stripes in an interferometer's beam pattern become narrower as the separation between the telescopes increases, and resolutions of 10^{-4} seconds of arc have been obtained. Such a resolution is ten thousand times better than the resolution of an optical telescope, but you should remember that this figure is the width of the stripes in a striped picture, not the sharpness of a detailed picture.

A VLB (Very Long Baseline) interferometer works in much the same way as the simple two-element interferometer described above, but the two telescopes are not connected by cables for they are thousands of miles apart. Instead, the signals are recorded on high-speed magnetic tape at each telescope, along with time ticks from an atomic clock, and then mixed later when the tapes are brought together by a computer.

The international aspects of VLB interferometry provide an interesting sidelight to this business. Most VLB work is done with very large telescopes, and few countries have more than one or two. In addition, the geographical size of a country limits the possible length of the baselines you can use. In the United States, for example, the longest baselines, or telescope separations, are three thousand miles. Thus a VLB astronomer can improve his results if he can develop a working relation with scientists from other countries. Perhaps the most noteworthy intercontinental experiment involved collaboration between American and Russian radio astronomers. These experiments presented a unique opportunity for achieving the highest resolution possible on the earth, as only the American and Russian telescopes were constructed well enough to observe at the shortest wavelengths, where the highest resolution can

FIGURE 8-6 Artist's conception of the Very Large Array. (NRAO photograph.)

be obtained. Aside from the scientific value of the experiment, it was very satisfying to see international cooperation working in the form of a research paper in the *Astrophysical Journal* with both American and Russian authors.

To give you an idea of how VLB interferometry works, I shall examine a particular quasar and show how a VLB astronomer extracts information on the quasar from the data he gathers.

A CLOSE LOOK AT A QUASAR: 4C 39.25

The quasar 4C 39.25, so named because it is source number 39.25 in the Fourth Cambridge catalogue of radio sources, is a seventeenth magnitude quasar with a redshift of 0.698. David Shaffer, in his thesis work at CalTech, found that this quasar had a well-determined structure at

2.8 centimeters wavelength, so I have selected it as an example of the use of VLB interferometry.

Three different antennas were used to observe this quasar. They were located at the Owens Valley Radio Observatory (OVRO) in Big Pine, California, the Harvard Radio Astronomy Station (HRAS) in Fort Davis, Texas, and the National Radio Astronomy Observatory (NRAO) in Green Bank, West Virginia. Each antenna observed the source from rise to set. Observations over an extended time period allow the antennas to assume different effective spacings, since the rotation of the earth changes the orientation of the baseline connecting the two antennas relative to the direction of the quasar. The magnetic tapes from each antenna were then brought together and combined, producing a "correlated flux density" for each pair of antennas. (The correlated flux density is the radio intensity measured when the two signals are combined.) The observations are shown as the vertical lines in Figure 8-7, where the length of the line indicates the error in the measurements.

But Figure 8-7 is not a picture of the quasar. It is just a "visibility curve." To figure out the structure of the quasar, Shaffer had to derive a model of the distribution of radio emission from the observations. He started with the reasonable assumption that the quasar consisted of some circular clouds with their radio brightness peaked at the center. He then calculated what the visibility curves of Figure 8-7 should look like if the model were correct, and adjusted the model to fit the data better. His final model produces the fits shown as lines in Figure 8-7, and its properties are described in his thesis:

> The resulting model is an unequal double with slightly resolved components. The double separation is 0.0020 ± .0001 seconds of arc in position angle 98° ± 3°. The two components contain all the flux and have an intensity ratio of ~4.5:1. The stronger component has a Gaussian width of 0.0004 seconds of arc and the weaker, ~0.0006 seconds of arc. These sizes are poorly determined, especially for the weaker component.[1]

I have depicted this model in Figure 8-8. The reason that this elaborate procedure of fitting a model to the visibility curve of Figure 8-7 is necessary is that the Very-Long-Baseline interferometers do not provide enough striped pictures of the radio sky so that the overlap can provide a nice neat map. Go back to Figure 8-5, and look at the right-hand panel again. Two pictures combined do not give you a very complete idea of where the radio noise is coming from. You need many, many striped pictures to unravel exactly what is happening. So do not think that Figure 8-8 is a well-determined map of the quasar 4C 39.25. It is just a sketch of a model. The observations are sufficient to show that this quasar must have at least two components, and there is no evidence for a more complex model. Yet there is no evidence against

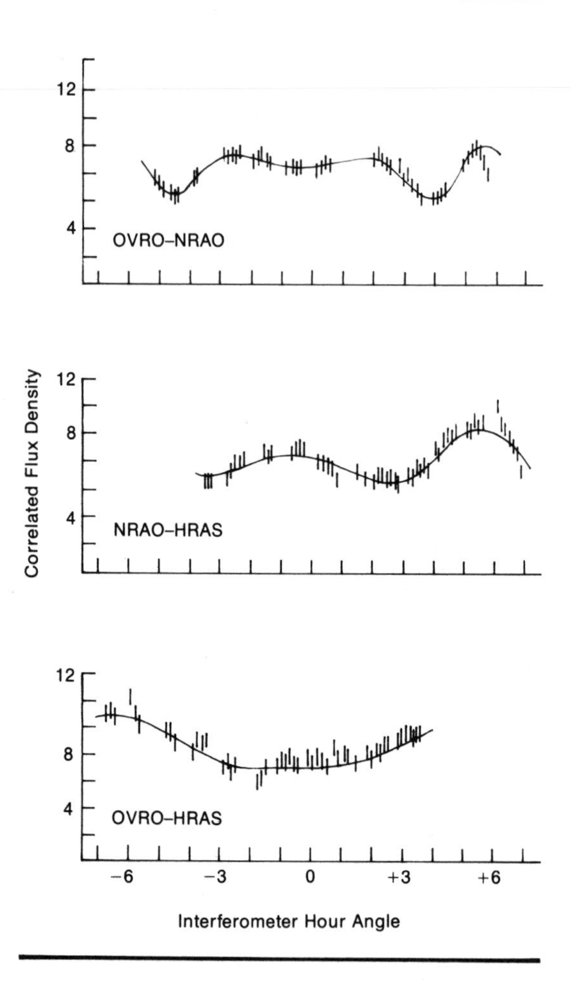

FIGURE 8-7 Very-Long-Baseline interferometer observations of the quasar 4C 39.25. The vertical lines give the correlated flux density, or the signal from the combination of the signals from each pair of antennas, as it varies as 4C 39.25 moves across the sky, changing the orientation of the interferometer baseline. The solid curves are the model fits of the model described in the text and depicted in Figure 8-8.

a more complex model either, and the source may well contain several components.

What can we learn from a map like Figure 8-8? We know that the radio emission is coming from a number of small clouds. At the 2790-megaparsec distance of 4C 39.25, these clouds are 3 and 5 parsecs across, according to the model, and they are 16 parsecs apart. The great achieve-

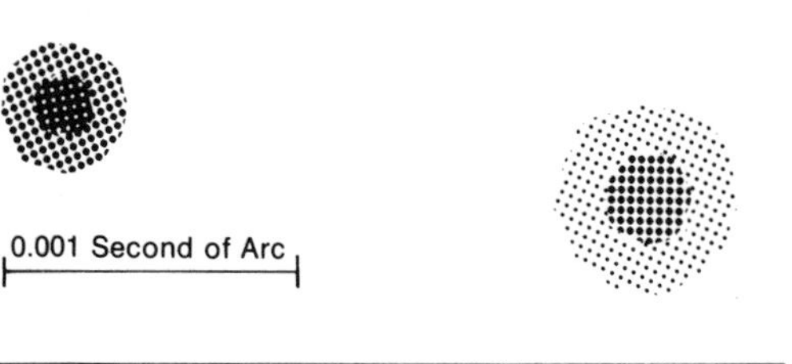

FIGURE 8-8 A sketch of the model of 4C 39.25 derived from the VLB observations of Figure 8-7.

ment of VLB interferometry is that it is possible to resolve such objects at such great distances. To do the same thing optically, an optical astronomer would have to be able to resolve an object seven inches across placed on the surface of the moon.

Very-Long-Baseline observations of other quasars show similar results. In general, the emission from quasars comes from a small number of clouds, each a few parsecs across. It is also generally true that the smallest clouds emit the higher-frequency photons. The number of clouds in each source varies, depending on the individual source. A few sources contain much larger clouds of radio radiation.

But these models leave many questions unanswered. It is quite puzzling that so much radiation comes from so small a volume of space. The total luminosity of 4C 39.25, for example, is about 10^{45} ergs/sec, or a hundred times the optical luminosity of a large galaxy. The numbers for other quasars are not very different. But all of this energy comes from a few clouds only a few parsecs across, ten thousand times smaller than a galaxy. How can such a gigantic amount of energy come from such a small cloud, and why does it come out in the form of radio waves?

The source of the radio-frequency radiation

Shortly after radio astronomy really began, radio astronomers realized that many of the sources that they were observing had nothing to do with stars. The radiation was not thermal—it did not come from heated objects. To emit strongly in the radio range and weakly in other parts of the electromagnetic spectrum, an object has to be unbelievably cool, a few Kelvins (you might wish to go back to Figure P-3). But if this radiation is not thermal, what is it?

In the 1950s, the Russian astronomer Iosif S. Shklovsky showed that the radio-frequency radiation from the Crab Nebula was a form of radiation called synchrotron radiation, coming from high-energy particles spiraling in a magnetic field. It was soon realized that most strong radio sources were emitting synchrotron radiation. We now know that all galaxies, including our own, are weak radio sources, and in our galaxy we have observed the raw materials needed for synchrotron emission: fast particles are seen as cosmic rays, and we have observed the weak magnetic field of the galaxy in several ways. Thus we have directly verified that the radio emission from our galaxy is synchrotron radiation. Before we examine a detailed model for the production of synchrotron radiation in quasars, it will be useful to know about the synchrotron process in a little more detail.

A synchrotron is a particle accelerator. Inside it, electrons or protons are accelerated so that they move at speeds close to the speed of light. Magnets in the synchrotron confine the motions of the fast particles so that they move in a circle. As these particles spiral around the circular path, they emit a ghostly light known as synchrotron radiation. This radiation is given off because the particles are forced to travel in circles, and the nature of the radiation depends on the energy of the spiraling particles. The higher the energy of a particle, the harder the magnetic field must work to hold it in its spiral and the higher the energy of the emitted radiation (Figure 8-9). High-energy radiation is also short-wavelength, high-frequency radiation, so that electrons of extremely high energy will give off x-rays, lower-energy electrons will give off optical radiation, and still-lower-energy electrons will give off radio radiation. All these electrons must move at speeds close to the speed of light in order to give off any synchrotron radiation, so even the low-energy electrons are moving quite fast.

Synchrotron radiation can be distinguished from other types of radiation because it is polarized. The electrons are only being accelerated around the magnetic field lines, not parallel to the field lines. Electromagnetic waves, whether they are light waves, radio waves, or something else, consist of electric and magnetic vibrations in space. In polarized radiation these vibrations operate only in one direction. Acceleration of the electron causes the electric field near the electron to vibrate in sympathy with the electron that is moving and in the same direction. Thus, in Figure 8-10, the electron is seen to move horizontally and the electric field to vibrate horizontally. When you investigate the synchrotron radiation with two antennas, one oriented horizontally and

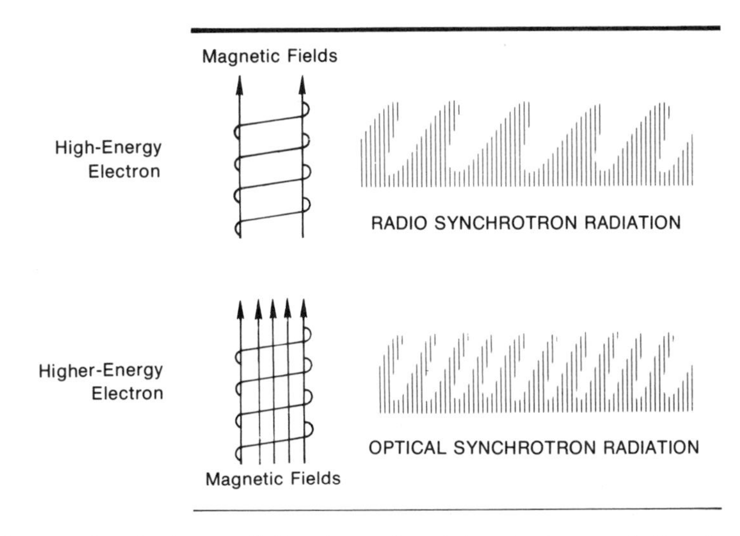

FIGURE 8-9 Synchrotron radiation is produced by electrons spiraling in a magnetic field.

the other oriented vertically, only the horizontally oriented antenna can pick up the polarized radiation. The antennas can be considered as pieces of wire with electrons free to move inside them, and only the electrons in the horizontal antenna can move in the same direction as the vibrations of the electric field. The vertical antenna will not pick up any radiation at all, because the electrons cannot move horizontally without moving out of the antenna. Another way of visualizing the situation is to think of the electrons in the two antennas as moving in the same directions as the electrons spiraling around the magnetic field. The synchrotron radiation is only a medium of communication between the two sets of electrons. Because the electrons in the vertically oriented wire antenna have no corresponding vertically moving electrons in the synchrotron radiation source, they do not move. Radiation that results from the acceleration of electrons in one direction only is polarized.

The polarization of the synchrotron radiation is the key to its origin. There are very few ways to produce polarized radiation in nature. The radiation from quasars is polarized, so one can conclude that the radiation from quasars is synchrotron radiation. In addition, the variation of the radio intensity with wavelength is consistent with what you expect from a synchrotron source.

You can now visualize how these small clouds in quasars emit radio waves. They contain large numbers of electrons moving close to the speed of light. (High-speed protons are not energetic enough to radiate much energy.) These electrons are spiraling around a fairly weak magnetic field, which causes them to emit radio waves by the synchrotron

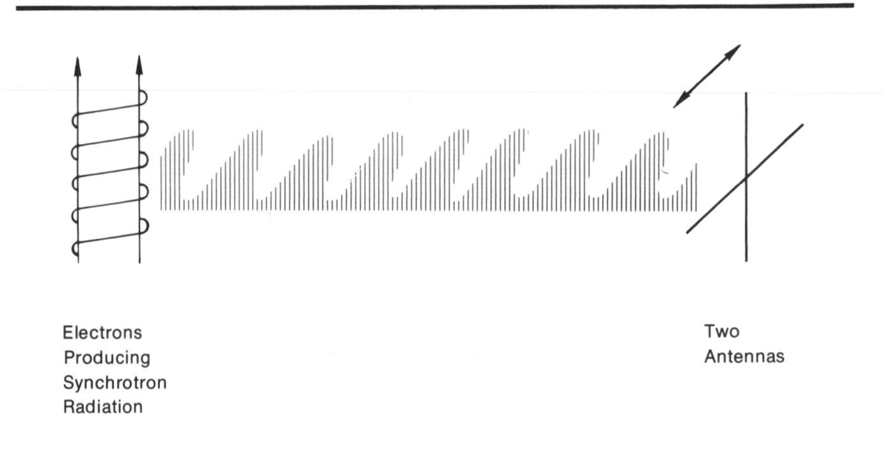

FIGURE 8-10 Synchrotron radiation is polarized.

process. The amount of energy contained in these fast electrons is tremendous, equal to the amount of energy radiated by an entire galaxy in millions of years.

This model sounds nice, but it is very qualitative. Before a reasonable comparison of model and reality can be made, we must try to be more definite. With no information about where these clouds come from, all we can do is to try to fit the experimental data to a synchrotron model of the kind described above. Such a fitting procedure begs the superimportant question of where these energetic electrons come from. That question underlies the deep mystery of quasars, so it is deferred until we know a little more about them.

Quasar radio observations: a more detailed model

These clouds of high-energy electrons are not static objects. In fact, it would be quite surprising if a parsec-sized collection of electrons moving at speeds close to the speed of light did not expand, move around, or at least do something besides just remain there radiating away. Since the quasars were first discovered, radio astronomers have been monitoring their brightness to see if these clouds have shown any signs of evolution. These signs could show up as variations in the total output of radiation from the quasar.

VARIABILITY

The brightest quasar, 3C 273, happens to be one of the most actively variable quasars in the radio range. Since it was first discovered in 1963, it increased steadily, until in 1966 it was almost three times as bright as it had been three years earlier. Then it declined for a while, losing about half of what it had gained, until it increased in brightness again in late 1967. Since then it has changed irregularly, increasing and then decreasing about every year or so.

These variations seem to be connected to changes in the wavelengths of the radiation emitted by quasars. When a quasar first brightens, it typically brightens first at short wavelengths, and then at longer wavelengths. This sort of behavior has not been observed too extensively, because uncovering such behavior requires a great many observations at different frequencies.

Such variability of quasars is consistent with the synchrotron model described earlier. The sudden increase in brightness occurs when a new cloud of high-energy electrons is ejected from the central source that is making them. At first, this cloud contains many electrons of very high energy, and these electrons radiate at high energies, or short wavelengths. Gradually, these high-energy electrons lose some of their energy, and the cloud begins to radiate at longer wavelengths. It also expands as time goes on, allowing the longer-wavelength radiation to escape from the cloud. It seems that new clouds are produced every year or so, and at any one time you can see three or four clouds in some of the more active sources.

For those quasars that vary measurably, calculations of the clouds' expansion rate and the way in which the spectrum changes with time are reasonably consistent with the observations. Remember, as we analyze quasars, it is important to have a quasar model that is not only qualitatively right but also quantitatively correct. Not only should our model describe general properties of the quasar (for example, synchrotron radiation correctly predicting that quasar radiation should be polarized), but the numbers must also work out correctly (for example, it should and does turn out that a cloud of electrons expanding on a time scale of about ten years adequately accounts for the changing wavelengths of radio radiation). This is especially important, because we are explaining data, not making predictions. When you can adjust your model to fit the data, it is important that the amount of data should exceed the amount of adjusting that you can do. While not all quasars are sufficiently well observed to be amenable to detailed model building, this picture of expanding clouds is fairly well confirmed for the slowly varying quasars.

MOVING CLOUDS

At one time it was widely reported that the clouds of high-speed electrons in quasars were not only expanding but also moving. There seemed to be evidence that these clouds were moving apart at speeds like ten times the speed of light. Such reports received considerable attention, since the idea of anything moving at ten times the speed of light activates numerous glands in the heads of speculative physicists. Some people argued that the quasars must be nearer than their redshifts indicated, so that the observed expansions would not be so fast. There was some wild theorizing about particles that travel faster than the speed of light, and some more sensible models were introduced, in which the clouds just seem to be expanding very fast. While it is not clear just what is happening, it now seems that this idea of faster-than-light expansion just came from the difficulty of interpreting the data from Very-Long-Baseline experiments.

One reason I spent a great deal of time explaining the workings of interferometry was to communicate the difficulties involved in interpreting the data. Remember that all that can be done with Very-Long-Baseline interferometers is model fitting. You take the data and fit a model to it. Some of the early models indicated that this faster-than-light expansion was taking place. As happens all too often, the uncertainties in the models and the whole details of the model procedure were not appreciated by non-VLB specialists, who took one look at the statement, "Model fits to the observations indicate that the clouds *may* be separating at superluminal, or faster-than-light, velocities," and unconsciously substituted the word *are* for the word *may*. It just happens that the appearance of a third very small radio source in a two-component quasar could make it look as though the other two components were moving apart very fast, because of the vagaries of the model-fitting procedure. It is not entirely clear how these VLB data should be interpreted, but it is now widely believed that the early data that apparently indicated superluminal velocities can be just as easily explained by variations in the intensity of individual components of the quasar.

Observations of quasars by radio astronomers which indicate that the following model explains the radio properties of quasars:

1. The radiation from quasars in the radio part of the electromagnetic spectrum comes from high-speed electrons, traveling at speeds close to the speed of light, spiraling in magnetic fields.

2. Very-Long-Baseline interferometer observations indicate that these electrons are localized in clouds, each a few parsecs across. Observa-

tions of the variations in the radio emission from quasars indicate that such clouds are produced every year or so in some of the more active quasars.

Yet you should keep in mind that these conclusions are based on model fits to the observations. The observers are ahead of the theoreticians in the quasar business, and these models are explanations. The whole synchrotron idea leaves the important consideration of the ultimate source of all of this energy undetermined. Yet there are other ways that a quasar puts out energy. We next turn our attention to these.

AN OPTICAL ASTRONOMER'S VIEW OF QUASARS

The picture of a quasar that has so far been built originates from the need to explain the quasar's great redshift and its radio luminosity. The radio emission comes from the synchrotron mechanism, which embraces a central energy source producing large clouds of relativistic electrons – electrons moving at speeds close to the speed of light. As these electrons gyrate around magnetic field lines, they produce prodigious amounts of synchrotron radiation.

Quasars emit a lot of optical radiation, too. The optical astronomer has no equivalent of Very-Long-Baseline Interferometry that can tell him the size of the clouds, but he has a marked advantage over his radio colleague in that much of the optical radiation is concentrated at certain wavelengths, producing emission lines from atoms. Analysis of these atomic emission lines can provide much information about the gas clouds or filaments that emit them – the size, density, temperature, and chemical composition of the clouds. A most puzzling feature of the optical spectrum is the existence of absorption lines in some quasars, indicating that there are cool clouds of gas between us and the quasar. Thus there are three different components of the optical radiation: the continuum, which extends over all wavelengths, the emission lines, and the absorption lines. Each will be considered in order.

The mysterious continuum

The radio synchrotron radiation (Chapter 8) is a form of continuum radiation, so named because it is not concentrated in any particular wavelength region but extends continuously over all wavelengths. Quasars emit such continuum radiation in many parts of the electromagnetic spectrum, as is shown in Figure 9-1. It has been observed in the infrared, the optical part of the spectrum, and in x-rays in the quasar 3C 273. We cannot tell whether the optical continuum is coming from the same clouds as the radio synchrotron radiation, for the resolution of optical telescopes is limited by turbulence in the earth's atmosphere to one

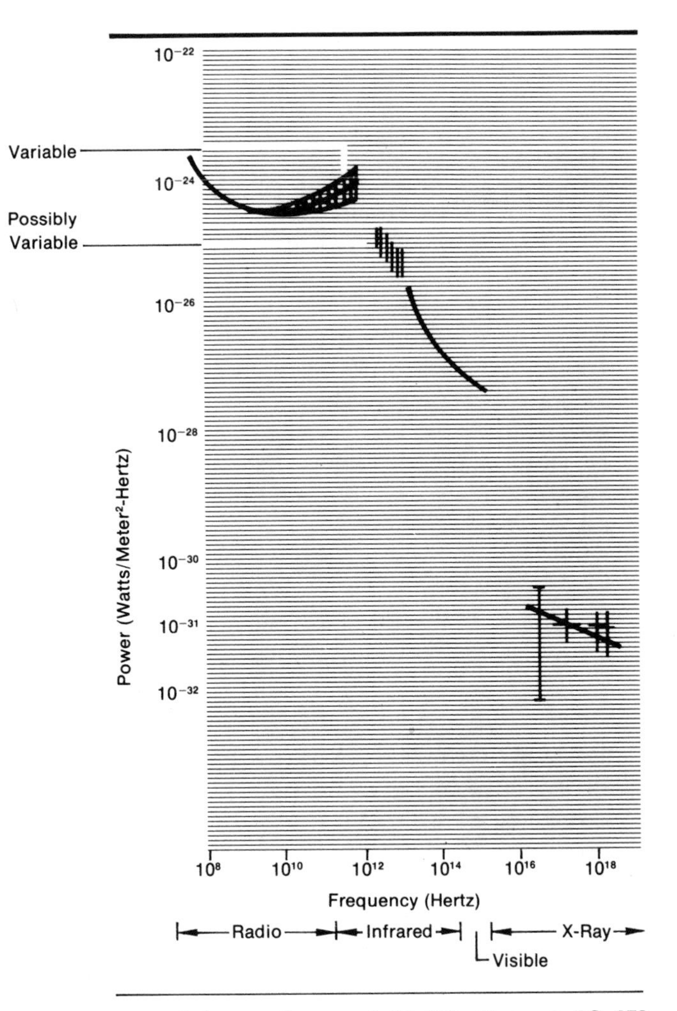

FIGURE 9-1 The spectrum of the continuum of 3C 273. Because 3C 273 is variable and the different measurements were taken at different times, these data represent an average spectrum, illustrating the gross features.[1]

second of arc, or about four kiloparsecs at the distance of 3C 273. If our galaxy were that far away, it could barely be resolved as a disk.

A few quasars, and 3C 273 is one of them, look nonstellar. Quasar 3C 273 has a faint jet protruding from it, and you may be able to detect this jet in Figure 7-8. The spectrum of this jet is continuous, and the jet is quite large. Yet the jet is much fainter than the rest of the quasar, so we shall concentrate on the spectrum of the main source.

Not all quasars have a spectrum like that of 3C 273. Some quasars do not emit radio-frequency radiation. At various times, these radio-quiet quasars were called interlopers, BSO's (blue stellar objects), or QSO's (quasistellar objects, rather than quasistellar sources). The name *quasar* is sufficient to cover these objects too; the radio-quiet quasars do not differ markedly in their optical properties from those that emit radio waves. The only quasar in which x-ray emission is intense enough to have been detected so far is 3C 273.

The intensity in the optical spectrum seems to match the radio spectrum fairly well, and it seems logical to conclude that the two forms of radiation come from the same process. On this interpretation, all of the continuous radiation is synchrotron radiation. The interpretation of optical radiation as synchrotron radiation is reinforced by observations of strong polarization of the optical radiation. In addition, the optical radiation has a spectral shape that resembles what you would expect from synchrotron radiation. The synchrotron model for optical radiation may face a problem when you consider the way that it varies.

BRIGHTNESS VARIATIONS

The optical radiation from quasars varies, just as the radio emission does. About four-fifths of all quasars vary in much the same way that 3C 273 does, brightening and fading over periods of years. Such behavior is very similar to the radio behavior of quasars, and can be interpreted in much the same way; a quasar brightens when a new cloud of relativistic high-energy electrons is produced, and then it fades as these electrons lose energy. The optical bursts are not the same bursts as the radio ones, since the optical and radio variations do not take place at the same time. However, the processes are similar.

About a fifth of all discovered quasars vary more dramatically. These quasars, called optically violent variables, change brightness rapidly. The best-studied one, 3C 446, has been known to double in brightness in a matter of a day or so. Such quasars may vary in the radio range as well, but the necessary observations have not been made.

This group of quasars comprising the optically violent variables presents a problem. The rapid variations of these quasars indicate that these objects are just a few light-days across. The volume of space responsible for the variation must receive the signal to vary in less than one day, and since no signal can travel faster than light, the object must be less than two light-days across. (Recall the discussion of Cygnus X-1 in Chapter 5; in that case, the variations were much faster, showing that the object producing them was very small.) If 3C 446 were any larger than two light-days, it would be like the brontosaurus—it just could not coordinate itself well enough to manage to cause a large gas cloud to start emitting in a short span of time.

What is wrong with having an object about two light-days across putting out so much radiation? It is difficult to construct a model that allows light to escape from the object without hitting an electron first. Synchrotron radiation is produced by electrons and magnetic fields, and radiation trying to escape from a small, dense cloud of relativistic electrons will hit the electrons before it can manage to escape. This difficulty is not faced by the synchrotron model for the radio emission, because the radio clouds are known through VLB measurements to be larger, and they vary much less rapidly than the optically violent variables. You encounter the size problem only when you try to explain the optical radiation from the rapidly varying sources.

When the size difficulty was first pointed out in 1966, some investigators thought that the problem was serious enough that the cosmological distances of quasars should be questioned. If you brought the quasars closer, they would not need to be as luminous and the size difficulties could be alleviated. Yet other ways around the problem have since been discovered that leave the quasars at great distances. If, for example, the cloud of electrons is expanding relativistically, at a speed close to the speed of light, you can produce a model that explains the rapid variations. While the situation is still unclear at present, most theorists feel that the small sizes suggested by the rapid light variations do not present a serious threat to the practicality of the synchrotron model. Problems arise only for the most rapidly varying sources, according to a recent treatment of the problem.[2] The size difficulty is really a quasar puzzle, not a fundamental problem.

THE INFRARED PUZZLE

Another puzzling aspect of quasar continuum radiation comes from observations of the infrared radiation. Look at Figure 9-1 again. Between the short-wavelength, high-frequency end of the radio spectrum and the visible there seems to be an excess of radiation. These observations, made mostly by Frank Low and his associates from a telescope mounted in a Learjet, indicate that 3C 273 is almost pathologically bright in this part of the spectrum. Quasar 3C 273 is putting out at least 90 percent of its energy as infrared radiation, emitting more than a hundred thousand times as much energy as our entire galaxy emits optically. Two questions arise about this infrared emission: Does it really exist? If so, what is it?

Making infrared observations is notoriously difficult because the earth's atmosphere is uncooperative. Infrared radiation is absorbed strongly by water vapor in the atmosphere, but if you can put your telescopes high enough you can obtain some useful data (see Figure P-4).

The near-infrared observations, which can be obtained from the ground, indicate some excess of radiation, but the real power seems to be coming in at a wavelength of some tens or hundreds of micrometers. Initially, the confidence that could be put in these infrared observations was fairly low, but when it was discovered that many other objects, such as the nucleus of our own galaxy, were emitting large amounts of energy too, it became a little more reasonable to accept the reality of the infrared radiation from quasars. These observations must stand the test of time. They must be repeated. At this stage, it does appear that there is an excess of infrared radiation, as shown by the spectrum of the continuum of 3C 273. What is not so clear is exactly how much of an excess there is.

Where does this infrared radiation come from? A logical idea would be to suppose that it is synchrotron radiation, in keeping with the synchrotron hypothesis for all of the continuum radiation. Yet this hypothesis encounters some difficulties. You need a prodigious quantity of electrons with energies of about one erg per electron. If their energy varied too much from this, you would be seeing visible or radio-frequency radiation. Somehow it seems strange that electrons would be preferentially produced with this amount of energy. Another idea is that the infrared radiation comes from dust grains in the vicinity of the quasar, which are heated and thereby emit infrared radiation. While the infrared radiation is polarized, according to recent data, either synchrotron or thermal emission from suitably aligned grains might be able to explain the polarization. We need more observations to decide whether the infrared radiation is synchrotron radiation, thermal radiation, or something else.

THE SYNCHROTRON MODEL

It may be that the continuous radiation of quasars, from x-ray to infrared, is all produced by the synchrotron process. There is good, solid evidence for this hypothesis in the case of the radio-frequency radiation, since the polarization and the distribution of energy with frequency seem to be consistent with the synchrotron model. The case in the optical and infrared parts of the spectrum is much less clear. It is probable that at least some, if not all, of the optical radiation is produced by the synchrotron process, since it is polarized. The infrared radiation is still a mystery, and tempting as it is to regard *all* continuous radiation as synchrotron radiation, we must remember that the synchrotron model is only a model, and that we must keep searching for more ways of comparing it with the real world situation. It must continue to meet the tests of reality. So far it has, more or less. There are a couple of puzzles yet to be solved: the cause of the rapid variations and the nature of the infrared emission. In addition, we have not yet found out where all of these high-energy electrons are coming from.

Hot gas clouds and emission lines

If quasars emitted only a continuous spectrum, it would be very difficult to verify their extragalactic nature. You would just see a smooth rainbow, and there would be no way to establish a redshift or a distance. Several such objects exist. Some have been shown by distance measurements to be nearby stars — white dwarfs. But some of these objects with continuous spectra vary in the same way that quasars do, and they have in many cases been identified with radio sources. They are called "Lacertids" in honor of the first one discovered, BL Lacertae, located in the constellation Lacerta (the Lizard).

Any bona fide quasar, however, must have emission lines in its spectrum, for a quasar is a starlike object with a large redshift, and you need emission lines to verify the existence of a redshift. Emission lines appear as vertical streaks in a spectrum (see Figure 9-2), indicating that

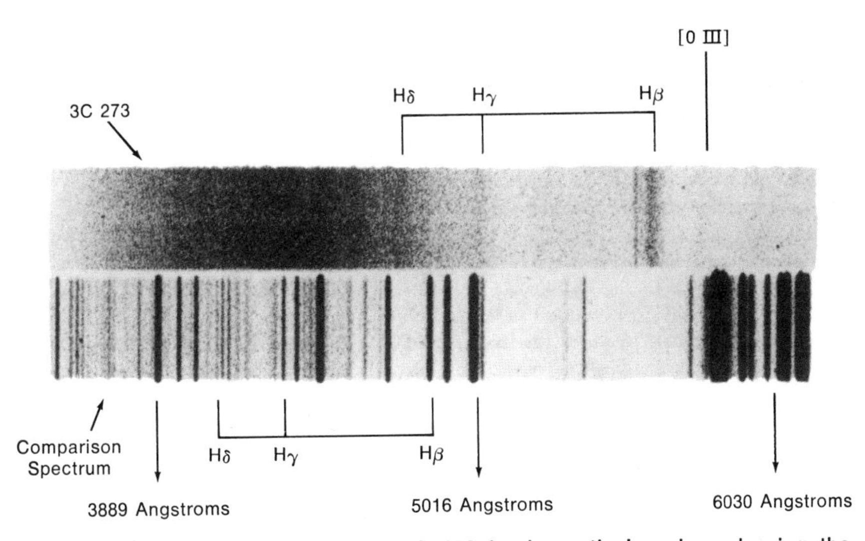

FIGURE 9-2 The spectrum of 3C 273 in the optical region, showing the emission lines of hydrogen (Hδ, Hγ, H_β, and [O III]) at the top. The bottom spectrum is a comparison spectrum of a helium-neon-argon lamp, exposed when the 3C 273 spectrum was obtained to give a wavelength scale. The figures on the bottom give the wavelengths, along with the unredshifted positions of the hydrogen lines. (Spectra from a Lick Observatory photograph.)

light emission is being concentrated at certain wavelengths. Recognition of a familiar pattern, such as the hydrogen spectrum in 3C 273, allows you to measure a redshift, as you then know what the wavelength of the emission lines should have been if the quasar were not moving away from us. (If you like this kind of activity, you might want to measure the wavelengths of the lines of the quasar spectrum in Figure 9-2, compare these with their rest wavelength, and see how close you come to the published value of $z = 0.158$. The rest wavelengths are 4101, 4340, 4861, and 5006 for Hδ, Hγ, Hβ, and [O III] respectively.)

SOURCES OF THE EMISSION LINES

What else can be determined about a quasar from these emission lines besides the redshift? Emission lines are an indication of the presence of a cloud of low-density gas that is heated by some source of high-energy radiation. A cloud of low-density gas is called a nebula (for example, see Figure 2-9). Emission lines detected anywhere in the universe betray the presence of low-density gas, whether in a nebula in our own galaxy, in an active galaxy, or in a quasar. Because emission-line spectra had been encountered before quasars were discovered, in nebulae in our galaxy, techniques for analyzing these types of spectra existed. They were soon applied to quasars in an effort to deduce properties of the regions in the quasar that were responsible for the emission lines.

To determine where these emission lines come from, consider an atom sitting around in space, some distance from a source of high-energy radiation (see Figure 9-3, frame 1). All atoms have energy levels. Their internal structure is such that an electron in the atom can exist only in certain energy states. It is like a staircase, where you can stand on the steps at only certain heights above the ground. When an electron moves between energy states, it gains or loses energy, just as you lose energy when you climb stairs and gain some energy of motion when you fall down stairs. In the case of the electron, the energy gained or lost is usually in the form of a photon. Our simplified atom has four energy levels, labeled by numbers.

The emission-line story starts when the atom is zapped by a high-energy photon (frame 2). The atom absorbs the photon, and the energy in the photon is used to kick the electron out of the atom, up to the shaded energy level marked "ion." Once the electron has escaped from the atom completely, it can have any amount of energy it wants to, so the "ion" energy level is not narrowly defined. What is left is an atom with one electron missing, or an ion, and a free electron wandering around in space.

The clouds of gas that produce the emission lines from quasars contain a lot of atoms sitting around in space. These atoms are illumi-

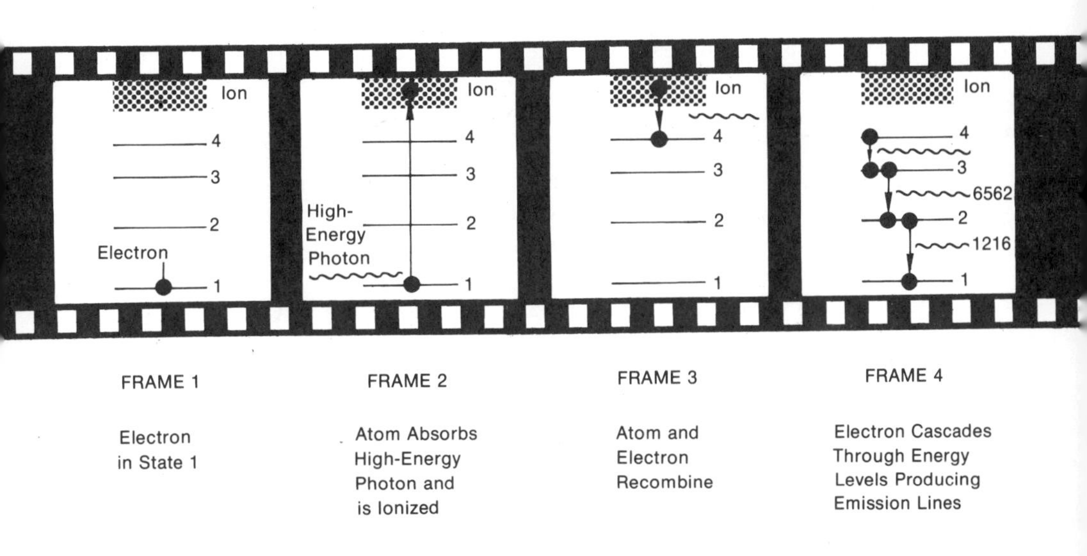

FIGURE 9-3 A movie showing how emission lines are made.

nated by a central source of high-energy radiation — the part of the quasar that is responsible for the optical and x-ray continuum. Some of this continuum radiation is energetic enough that it can be absorbed by one of the atoms in the cloud, as in frame 2 of Figure 9-3. The electron escapes from the atom and flies off into space.

Eventually this electron will collide with an atom, as this gas cloud is filled with electrons and ions. The atom and electron recombine (frame 3), as they get back together again in what should be their natural, neutral state. However, the electron is not necessarily captured directly into the lowest energy level (No. 1 in this case); often it is captured into a higher energy state, No. 4 for instance. When the electron and atom recombine, the electron loses some of its energy. Because energy is neither created nor destroyed, a photon is emitted and carries away this extra energy that the electron gave up.

But will the electron sit around in level 4 forever? No, because electrons are lazy and seek the lowest possible energy state. The electron now cascades downward through the different energy levels of the atom (frame 4). It can drop through the levels sequentially, as shown here, or it can skip over a few levels, just as a person going down a flight of stairs can take the steps one, two, or three at a time. Eventually the electron ends up in level 1 again, ready to be ionized by another high-energy photon and start the sequence over.

The emission lines are produced, shown in frame 4, as the electron cascades downwards through the atom's energy levels. Since energy is neither created nor destroyed, the electron must get rid of its energy as it drops down these levels. As it goes from, say, level 3 to level 2, it loses an amount of energy equal to the energy difference between levels 3 and 2, or 1.89 electron-volts if this is a hydrogen atom. This energy is given off in the form of an emitted photon, which eventually may find its way to planet Earth and our telescopes. When this photon leaves the quasar, it has an energy exactly equal to the energy difference between levels 3 and 2. Since photon energy is related to photon wavelength, this photon has a wavelength of 6562.8 angstroms when it leaves the quasar.

Any atom of a given element has a distinct pattern of energy levels, different from the pattern of any other atom or any other ion. This pattern of energy levels shows up in the pattern of energies of photons emitted as the electrons cascade downward through the atom. Hydrogen, for example, will produce emitted photons with wavelengths of 6562.8 angstroms (level 3 to level 2, as shown), 4861.3, 4340.5, and 4101.7 angstroms (levels 4–2, 5–2, and 6–2 respectively), and 1216 angstroms for a jump between levels 2 and 1. (This list does not exhaust the hydrogen spectrum.) Because hydrogen is the most common element in the universe, this pattern of emission lines produced by descents to the second level is familiar to most astronomers: 6562.8, 4861.3, 4340.5, and 4101.7 angstroms. The last three of these are visible in Figure 9-2 as the lines $H\beta$, $H\gamma$, and $H\delta$.

Whenever you see an emission-line spectrum from an astronomical object, you know that the object contains a cloud of low-density gas that is subject to ionizing radiation. There is no other way to produce an emission-line spectrum. A quasar contains a source of ionizing radiation – the clouds of high-energy electrons that are producing synchrotron radiation. The continuum is being produced at all wavelengths from radio waves to x-rays. To ionize hydrogen atoms, ultraviolet photons are needed. There are probably plenty of such photons around as a result of synchrotron radiation.

SPECTRA OF QUASARS

Completing our outline, we can now add an emission-line region to our picture of a quasar. Gas clouds surround a central source of continuum radiation. This gas acts as a converter, absorbing ultraviolet radiation from the central source, ionizing its atoms, and then producing emission lines. It is generally believed that this gas is concentrated in clumps or filaments (see Figure 9-4). The reason we believe the gas is concentrated in clumps is that some of the ionizing radiation escapes to the outside and can be seen through our telescopes when it is redshifted into the visible spectrum in the quasars with the highest redshifts. Furthermore, it is often thought that quasars are a kind of super–Crab Nebula (Chapter 3), and that the Crab contains filaments too. In a typical quasar, the total amount of ionized gas in all the filaments is about 10^5 solar masses, and it occupies a total volume of about 1000 cubic parsecs.

We should not stop here with the process of constructing a generalized model for the emission-line region of a quasar. The schematic model depicted in Figure 9-4 must be made more quantitative and applied to particular quasars. We must enter the normal science process of making detailed, not just generalized, models of particular objects and seeing whether they fit the real world of observations. You must consider such questions as, Can you produce the right kind of emission spectrum? Are the relative intensities of different emission lines in reasonable agreement with what the model says they should be? Are the atoms whose spectra are found in quasars the kinds of atoms that you would expect to find there? Figure 9-4 is just a skeleton; you need to add some flesh to it. This process is going on now; there is enough data in the emission-line spectra of quasars that the data is amenable to this kind of detailed model building. Studies of the intensities of emission lines in quasars are particularly satisfying in that we can get away from generalized arguments. Theoreticians need not be content with explaining general trends; they can calculate emission-line strengths from models of hot gas clouds subject to ionizing radiation. These calculations are now going

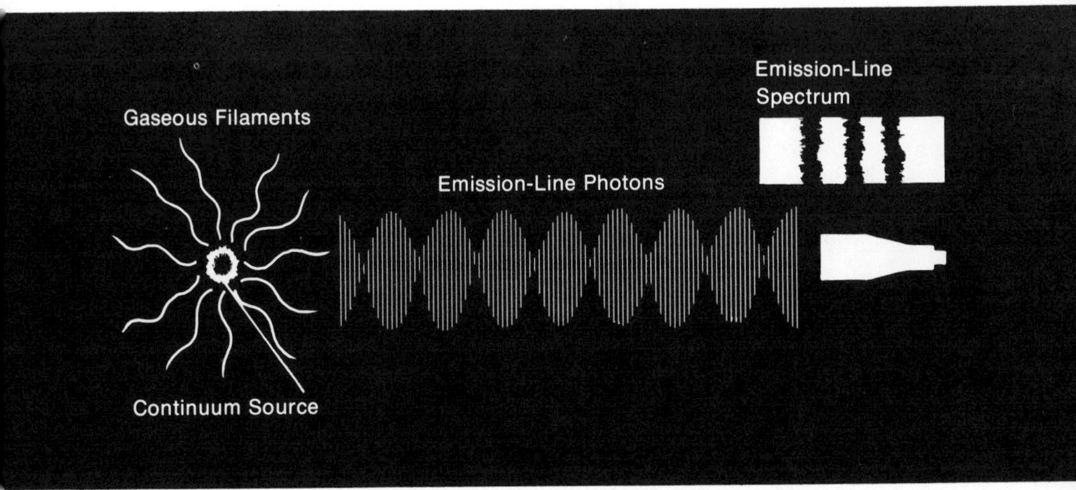

FIGURE 9-4 A generalized model for the regions of a quasar that are producing the emission lines.

on and the model agrees with the observations.[3] Hydrogen, some helium, carbon, oxygen, neon, magnesium, and argon are observed in quasar spectra, and model analyses of these spectra indicate that the relative abundance of these different elements is, as far as we can tell, entirely normal. The model agrees.

Motions in Quasars

One striking characteristic of the emission lines in quasar spectra is their breadth. If the filaments of gas in the quasar were at rest relative to each other, someone looking at the quasar from a distance would see every emission-line photon produced by a hydrogen atom cascading from level 3 to level 2 appear at the same wavelength, 6562.8 angstroms. The atoms within the filaments are moving, however, because the filament is hot; this motion would cause these emission-line photons to be Doppler-shifted, some to the red, and some to the blue. The shifts produced by these atomic motions are due to the heat of the gas in the clouds causing the recombination radiation. Yet these thermal Doppler shifts are quite small. For the temperatures derived for these gas clouds from analysis of the emission-line spectra, the hydrogen lines would be about 0.1 angstrom wide on the spectrum as a result of thermal broadening. In fact, the hydrogen lines are smeared out, some 50–100 angstroms wide in the spectrum.

The tremendous breadth of these emission lines indicates that the emitting atoms are moving relative to each other at great speeds, some thousands of kilometers per second. Some of the emitting atoms are moving away from the observer, and some are moving towards the observer, relative to the center of the quasar. As a result some emission-line photons are redshifted and some are blueshifted, relative to the average photon. The cause and consequences of this motion have not been fully understood. Is it due to rotation of the quasar? Are the filaments exploding away from the central source? Are these motions random or systematic? Whatever they are, they are indications that the filaments producing the emission lines are in rapid, even violent motion.

ABSORPTION LINES: COOL CLOUDS?

Another indication that quasars might contain rapidly moving clouds of gas is the presence of absorption lines in their spectra. Absorption lines show up as gaps in the spectrum, indicating that there is less radiation emitted at that particular wavelength than at neighboring wavelengths. These absorption lines are less redshifted than the spectrum of the quasar as given by its emission lines.

Absorption lines are familiar features of the spectra of stars, and they are formed in the same way in quasar spectra. As the quasar spectrum passes through a cool gas cloud between us and the quasar, a hydrogen atom, for instance, will steal a photon of 1216-angstrom wavelength from the radiation from the quasar, and this atom will move its electron to a higher level. The absence of this stolen photon will show up as a gap in the quasar spectrum. (You might wish to refer to the discussion of stellar spectra in the Preliminary section.) The wavelength of the gap is not shifted as much as the wavelengths in the emission-line spectrum of the quasar itself, indicating that the absorption-line cloud is not moving away from us as fast as the quasar is. In all quasars in which absorption lines have been seen, the redshift of the absorption lines is less than the redshift of the emission lines.

Where are these clouds of cool gas? We know that these clouds are cool because the absorption lines are quite narrow, indicating that the atoms in the absorption-line cloud are not moving very rapidly relative to each other. Most astronomers think that these cool clouds are near to the quasars, and that the difference in redshifts comes from a velocity difference between the quasar and the cloud. Presumably the cloud was ejected from the quasar at one time. It is also possible that the absorption lines come from dead or unseen galaxies between us and the quasar. If one could show that this was the case, one could argue that the quasars are more distant than these dead intervening galaxies, or that the quasars are at cosmological distances. Yet all of these are vague possibilities, unsupported and uncontradicted by theoretical models. We really do not yet understand the absorption lines in quasars.

Because the optical spectrum of a quasar has emission-line features, you can learn a good deal about a quasar from interpreting it. The optical spectrum contains three features: an underlying continuum, not concentrated at any particular wavelength, emission lines, and absorption lines. At least some of the continuum is probably synchrotron radiation from small clouds of high-energy electrons close to the center of the quasar. The emission line spectrum comes from filaments of hot gas surrounding this source. This hot gas is subject to radiation that ionizes its atoms, producing the emission lines as recombination radiation. The smearing of the emission lines in the spectrum indicates that these filaments are moving at very high velocities. The absorption lines come from clouds of gas between us and the quasar, but whether these clouds are part of the quasar or not is quite confused at this time.

One can put all of this information together and derive a schematic model for a quasar, shown in Figure 9-5. (If you think this looks a little like the Crab Nebula, you're right. The Crab Nebula is the astrophysicist's model for most energetic phenomena!) In the center is an energy source, whose mysteries we have not yet begun to fathom. It does all the work of generating energy, and it generates this energy in the form of high-energy electrons. The rest of the quasar consists of various clouds. The inner set of clouds produces the optical, infrared, and x-ray continuum radiation, along with the ultraviolet continuum that we cannot see too much of because it is ionizing the filaments and producing emission lines. This radiation is probably produced at least in part by the synchrotron process. The clouds responsible for the optical emission are, in some cases, a few light-days across – very small compared with the rest of the quasar. Very-Long-Baseline observations of the clouds producing the radio emission indicate that these clouds are somewhat larger – a few parsecs across.

The emission line photons are produced by the filaments. The net volume of these filaments is some 1000 cubic parsecs and they contain

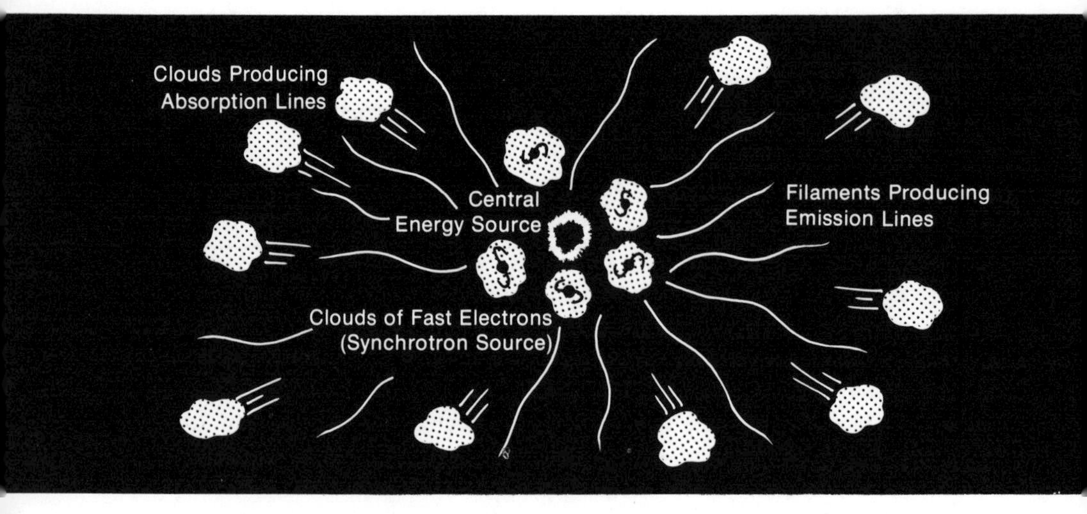

FIGURE 9-5 Generalized model of a quasar (see text).

about 10^5 solar masses of ionized gas. They are probably the best understood part of the quasar. Still further out may be the clouds producing the absorption lines, but we are not really sure where they are.

We have still not yet attacked the nature of the central energy source. We can learn some more about this remarkable object before theorizing about its nature, however, if we look at the active galaxies, which are objects that are in some respects similar to the quasars.

ACTIVE GALAXIES

A galaxy is basically a large collection of stars. It seems that the light from the galaxy as a whole should simply be the total of the light from every individual star, and that the evolution of a galaxy could be completely understood if you knew how and where star formation and star death were taking place. This view is a nonviolent one — that galaxies are nothing more than gigantic star clusters, and that their evolution proceeds at the slow, stately pace of stellar evolution.

In the 1960s, evidence accumulated that quite a few galaxies are not calm and quiet places. Some galaxies showed evidence of large masses, millions of suns, of ionized gas moving very rapidly, as though shot out of the galactic nucleus by an explosion. Others radiated vast quantities of synchrotron radiation, showing that some violent event occurred that produced prodigious quantities of high-energy electrons. A galaxy that is doing any one of these things is termed an *active galaxy,* since it is doing something more than just producing starlight. The diverse types of galactic activity make a dazzling display and are probably connected with the quasar phenomenon.

The quasar model summarized at the end of the last chapter is tantalizingly incomplete. Analysis of the different types of radiation — continuum, emission lines, and absorption lines — presents a quasar as an object containing several different types of gas clouds surrounding a central energy source. (Recall Figure 9-5.) These clouds produce the different types of radiation. Some contain high-energy electrons and produce synchrotron radiation, some contain hot gas and produce emission lines, and some produce absorption lines. But two basic questions remain: What is the relation of the different types of clouds — do the different types of clouds evolve from each other, or are they produced by fundamentally different mechanisms? Where does all this energy come from in the first place? Many details of the cloud picture still remain to be worked out, and it may be that these details yet to come will provide hints to the answers to the two main questions.

Much insight into the quasar phenomenon can come from quasar analogues, which though less energetic are closer to us: the active galaxies. The active galaxies are close enough to the earth that we can resolve the distinct clouds in each galaxy, thereby obtaining a better picture of what is happening. The quasars are so far away that they just look like point-sources of light, and provide much less information.

Signs of activity in galaxies

An active galaxy does something more than just produce starlight. This activity is exhibited in a bewildering variety of ways: synchrotron radiation, broad emission lines, peculiar optical appearance in photographs, and infrared radiation. Nearly all galaxies, including the Milky Way, are active; it is only the irregular galaxies and the loosely wound spirals that seem to contain nothing more than shining stars. This activity tends to originate in the nucleus of each galaxy, and since irregulars have no nucleus it is not surprising that they are not active. How is the activity displayed?

OPTICAL APPEARANCE

The most active galaxies frequently look peculiar in photographs. You might want to go back to Chapter 7 and look at the photographs of ordinary galaxies, comparing them with the peculiar galaxies of this section. A number of different peculiarities are evident.

Two active galaxies and one quasar are associated with jets. Compare the photographs of 3C 273 (Figure 7-8), Messier 87 (Figure 10-3), and the Seyfert galaxy NGC 1275 (Figure 10-7). Each one has a jetlike protuberance. It is curious that all three of these objects are x-ray sources. Jets are seen only in active galaxies.

The only analogue of galactic activity that we see in our own galaxy, the Crab Nebula, displays many of the same characteristics as active galaxies, only on a much smaller scale. The Crab Nebula was named because through a medium-sized telescope the filamentary structure looks like crab legs. This filamentary structure is also characteristic of active galaxies; compare the photographs of Messier 82 (Figure 10-2) and NGC 1275 (Figure 10-7) to the Crab Nebula (Figure 3-1).

Some active galaxies show prominent dust lanes. The prototype here is NGC 5128, the radio source Centaurus A (Figure 10-5); Messier 82 contains much dust, portrayed to some extent in Figure 10-2.

Nearly all active galaxies have bright nuclei. It is difficult to pick out these nuclei on photographs, which must be overexposed so that the rest of the galaxy can be photographed. Figure 10-6 shows some N-galaxies, which have "brilliant, star-like nuclei containing most of the luminosity of the system."[1] I shall return to these later.

The different peculiarities of active galaxies are of great use to astronomers seeking to discover new ones. You look at photographs of the sky and make a list of objects that look peculiar and seem worthy of closer attention. The rewards of search work are that your name is attached to this list of peculiar objects, and these objects bear your name. A disadvantage of this historical process is that a bewildering variety of terminology has crept into the literature, presenting a confusing array of terms describing different types of active galaxies. I have supplied a guide to this terminology in the Appendix to this chapter, to help with the literature.

Yet optical peculiarities, useful as they may be to people trying to discover active galaxies, do not tell us very much about what is going on in these galaxies. How can you investigate such activity once you have discovered it?

SPECTROSCOPIC CHARACTERISTICS

Activity in galaxies is most clearly shown in a spectrum of the galaxy in question. A galaxy that is radiating starlight produces a spectrum that looks like the spectrum of a star: a continuous band of light with a few dark lines crossing it (recall Chapter 5 and the Preliminary section). Active galaxies generally show emission lines in their spectra. Where there are emission lines there are clouds of hot gas, and this gas is ionized by some source of energetic ultraviolet radiation.

SYNCHROTRON RADIATION

Where you find synchrotron radiation, there is some object that can accelerate electrons to the high speeds necessary to produce it. Synchrotron radiation can show up at all wavelengths, but it is most obvious at those wavelengths where stars do not radiate very much: x-ray, infrared, and especially radio. Many active galaxies have been discovered because they were counterparts to radio sources.

CLASSIFICATION OF ACTIVE GALAXIES

With such a wide variety of phenomena showing activity in galaxies, how can you bring some order to the observations so you can even begin to attack the problem? Many classification schemes exist, but most are based on the optical peculiarities of the galaxies rather than on any underlying properties. The scheme I use to order the discussion in this chapter is based on the relative scale of galactic activity.

Galaxies can do two things: they can contain stars that shine or they can contain more exciting objects that explode, produce synchrotron radiation, or do something else that is violent. Most galaxies show some signs of violent activity, which can be classified by examining the relative amount of effort that the galaxy puts into nonviolent (stellar) and violent radiative processes. It is difficult to attach any specific numbers, like a ratio, to these two processes, as much of the violent activity comes out in the infrared, where observations of exactly how much energy is emitted are difficult to obtain. But the order of different types of objects seems to be reasonably well established. Galaxies like the Milky Way put out about 1 percent of their total luminosity in the form of nonstellar radiation, since most of their radiation is just starlight. At the other end of the classification scheme are the quasars, from which we see no ordinary starlight at all. If the quasars contained as many stars as giant elliptical galaxies do, their nonstellar luminosity would be hundreds or thousands of times greater than their stellar luminosity. We start with normal spiral galaxies, on the inactive end of the scale; and the nearest spiral galaxy — and the one we know best — is the one in which we dwell, the Milky Way.

The core of the Milky Way

The center of our galaxy may be a miniature quasar. While the nonstellar luminosity is only 1 percent of the total stellar luminosity of the galaxy, all the events that we see in quasars are taking place on a reduced scale. Since we are only ten kiloparsecs away from the galactic core and 900,000 kiloparsecs away from the nearest bona fide quasar, 3C 273, we should be able to view this activity in more detail.

Unfortunately our view of the galactic center is limited, as our galaxy contains many dust clouds that block our view. What looks like the galactic center in Figure 7-1 is really only part of the way towards the center. Light cannot penetrate these clouds, but infrared and radio-frequency radiation can, so a much better picture of the galactic center has been obtained in the last few years with the development of infrared and radio astronomy. If we wish to obtain an optical view of the center of our galaxy, we can turn our telescopes onto our nearest neighbor, the Andromeda galaxy, or M 31. This galaxy is quite similar to our galaxy in size and shape.

Infrared radiation can penetrate the blanket of interstellar clouds and show us a picture of the star distribution near the galactic center. The star concentrations unveiled by this radiation are quite similar to the star concentrations seen visually in the Andromeda galaxy, reinforcing the parallel between the two systems. The galactic core contains a tremendous concentration of stars, about 10^5 to 10^6 stars per cubic parsec. Stars at the core are sufficiently close to each other that an encounter between two stars would knock any planets out of their orbit. Most of these stars are older ones, like the sun; there are few young, blue stars.

These results come from a composite picture developed from infrared observations of our own galaxy and photographs in visible light of the core of M 31. The star concentration extends over a region a few hundred parsecs across – the nuclear disk.

At the center of this nuclear disk is the first radio source ever discovered – Sagittarius A. Sagittarius A is about ten parsecs across and is a powerful source of synchrotron radiation. The central region of Sagittarius A contains an even stronger, very small source of infrared radiation, whose nature is unknown at this time. The radio power emitted by Sagittarius A is 10^{37} ergs/sec – not large when compared with the 10^{45} ergs/sec radio luminosity of some quasars, but still 10^{4} times the luminosity of the sun.

The nuclear disk surrounding Sagittarius A contains a number of components whose interrelations are not known. This disk contains the central concentration of stars, but it is also emitting radiation at x-ray and far-infrared wavelengths. The infrared radiation may be synchrotron radiation, but it seems more likely at present that it comes from dust grains heated by the light from the central stars. The disk contains a number of clouds of hot gas, which is ionized by some source of ultraviolet radiation. Also present in the disk are numerous very cool clouds, containing some molecules of water, ammonia, carbon monoxide, alcohol, formaldehyde, and other substances. These molecular clouds must contain a good deal of dust to absorb the ultraviolet radiation, which would otherwise destroy the molecules. There is some evidence that the cool molecular clouds form a ring around the center of the galaxy, 300 parsecs in radius.

Further away from the core, two arms of the galaxy are expanding away from it. One arm is on the far side of the nucleus, moving with a speed of 135 kilometers per second away from the core. The other, on the near side, moves towards us with a velocity of 53 kilometers per second. It is not yet clear where the energy came from to power the motion of these expanding arms. One view has it that the galactic core was rocked by a stupendous explosion about ten million years ago.

A tentative picture of the nuclear region of the Milky Way galaxy is shown in Figure 10-1. Our galaxy contains many of the same things found in quasars: clouds of relativistic electrons producing synchrotron radiation, moving clouds of hot gas, an infrared source, and cool gas clouds. But we still do not know how it all fits together. The view described in this section is phenomenological, not evolutionary; we know what is there but we do not yet know how it came to be. But the galactic core is close enough that we can perceive the spatial relationships between the different sites of activity. We can see that the synchrotron source, Sagittarius A, is small and at the center, while the near-infrared radiation comes from a much larger disk. With all this information we

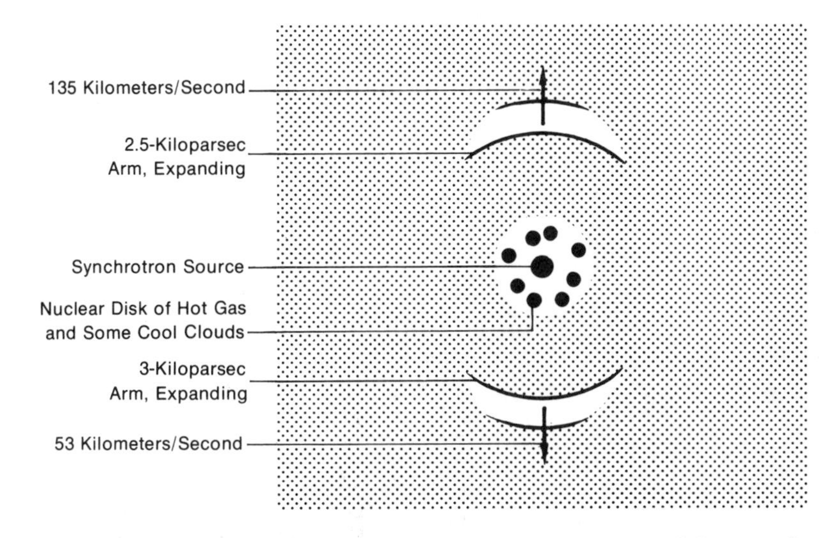

FIGURE 10-1 A schematic depiction of some of the features of the galactic center, not drawn to scale (see text for details).

should be able to unravel the presently tangled picture in the not too distant future, and any insight into the galactic core will help us understand the more intense activity of quasars.

Most spiral galaxies seem to show somewhat similar signs of activity. Gas shows up in the form of emission lines. It seems that wherever there is a galactic nucleus, there is some action. Some galaxies, such as the Magellanic Clouds, have no nuclei and thus no synchrotron source or expanding arms. Dull places, those. But we can move up on the activity scale to a very strange spiral galaxy, Messier 82.

Messier 82: a hyperactive "normal" galaxy

Messier 82 is the nearest galaxy to show activity on a scale comparable to the stellar luminosity of an average spiral galaxy. Figure 10-2 shows its irregular, Crab Nebula–like apearance in a photograph taken in the light of an emission line, so the only radiation that registers on the photograph is from clouds of ionized gas.

The peculiarities of Messier 82 first became evident when C. Roger Lynds noticed that it contained a weak radio source at its center. In the early 1960s, the power of the large telescopes at Lick Observatory and Mount Palomar was turned on this object, and it was discovered that much of the luminous energy from the galaxy was in the form of emission

FIGURE 10-2 Messier 82 in the light of the H_α emission line. (Hale Observatories.)

lines, indicating that there are vast quantities of hot gas subject to high-energy, ionizing radiation. Doppler shifts in these emission lines seemed to indicate that the filaments, shown in Figure 10-2, were expanding away from the center at speeds of 1000 kilometers per second. It was thought that M 82 was rocked by a violent explosion that tore the galaxy apart. Recently, infrared telescopes were turned on the galaxy, indicating that somewhere around 10^{44} ergs/sec of energy were being emitted as infrared radiation. It is clear that something very unusual is occurring in M 82, but it is not clear exactly what.

In the last few years, Alan Solinger of MIT suggested that maybe M 82 is not exploding at all. His idea is that the visible filaments are really made of dust, which scatters the emission-line photons that are produced somewhere else. This dust is moving, but it does not need to be moving very fast to explain the Doppler shifts of the emission lines. If the dust model is correct, then the connection between M 82 and the quasar phenomenon is not so direct.

We are a long way from unraveling the mystery of Messier 82. Is it exploding, or is it not exploding? Why are the outer filaments bluer than the inner filaments? Where is the energy coming from to produce the emission lines? What is the source of the infrared radiation? As Allan Sandage and Natarjan Visvanathan, two assiduous investigators of this puzzling galaxy, confessed in a recent article on M 82, "The galaxy remains a mystery to us."[2] Nevertheless, something exciting is going on there. This galaxy is a potential link between our own, relatively non-violent Milky Way and some of the galaxies higher up in the hierarchy of activity.

Radio galaxies

Normal spirals and M 82 are not examples of especially energetic activity in galaxies. Normal spirals put out 99 percent of their energy as starlight, and while M 82 is still quite mysterious, the overall energy level of the phenomena in that galaxy is still fairly low. The next step up the scale involves the radio galaxies.

There are two principal types of radio galaxy. One type contains compact radio sources, which are in some cases less than a parsec across. The luminosities of these compact sources are considerably less than quasar luminosities. The extended radio galaxies, considerably more common, put out most of their radio emission from extended clouds of high-energy electrons, which are often hundreds of kiloparsecs away from the galaxy that seems to have produced them. (For comparison, the Magellanic Clouds are about 60 kiloparsecs away from the Milky Way.) It is not yet clear what the association between the compact sources and the extended sources is. Compact sources are generally associated with some other form of galactic activity (brilliant nuclei, emission lines, or infrared emission), while the extended sources are not necessarily so linked. While the observations are far from complete, it seems that quasars always have a compact component and that sometimes they have an extended component.

A COMPACT RADIO SOURCE: MESSIER 87

Messier 87 is the second closest strong radio galaxy, and it dominates the Virgo cluster of galaxies (Figure 10-3). It is an important key to our understanding of active galaxies, for at its distance of 15 megaparsecs it is close enough that we can see in detail just where the activity is concentrated. Furthermore, it is one more step up the scale of activity.

The radio emission from M 87, all synchrotron radiation, comes from three distinct regions. Most of the emission comes from a large halo about 60 kiloparsecs across, roughly the same size as the visible galaxy. A remarkably compact core, only 2.5 light-months across also produces radio noise. This core is the most compact radio source known to be connected with the radio galaxy phenomenon. The optical jet contains several radio sources.

The most obvious sign of optical activity is the jet extending from

FIGURE 10-3
Messier 87, showing the jet. (Hale Observatories.)

the nucleus of M 87, visible on the short-exposure photograph of Figure 10-3. On heavily exposed photographs, a counterjet extending in the opposite direction is seen. This jet is a string of highly condensed regions, each of which emits a lot of polarized, presumably synchrotron radiation. Since a high-energy electron could not travel from one end of the jet to the other without losing its energy, it is generally supposed that the knots in the jet surround individual sources of high-energy electrons.

Most of the nonstellar radiation produced by M 87 is in the form of x-rays. Its x-ray luminosity is about 10^{43} ergs/sec. It is not yet known whether these x-rays are produced in the jet, in the halo, in the compact central source, or in all three locations, since x-ray telescopes cannot produce such high-resolution photographs. These x-rays are probably synchrotron radiation as well.

It is not yet clear how these diverse indications of activity in M 87 are to be fitted into a complete picture. Presumably something at the core of this galaxy is producing large quantities of high-energy electrons. Some of these electrons are shot out in the direction of the jet, while others end up in the halo. Emission lines are observed in parts of M 87, indicating that the galaxy contains sources of ultraviolet radiation also. Some energy sources are located along the jet. Future observations will probably tell us more about where this galaxy generates its energy, and once we know where the energy is generated, we may be able to start asking how it is generated. Messier 87 is the prototype of one kind of radio galaxy: the core-halo galaxy. There is also another type of radio galaxy, the extended radio galaxy.

CYGNUS A, AN EXTENDED RADIO GALAXY

Cygnus A, one of the strongest radio sources in the sky, is an excellent example of a double radio galaxy (see Figure 10-4). (The strongest radio sources are often referred to by the name of the constellation in which they are located. Thus Messier 87 is also called Virgo A.) Its radio emission, 1.2×10^{45} ergs/sec, comes from two enormous

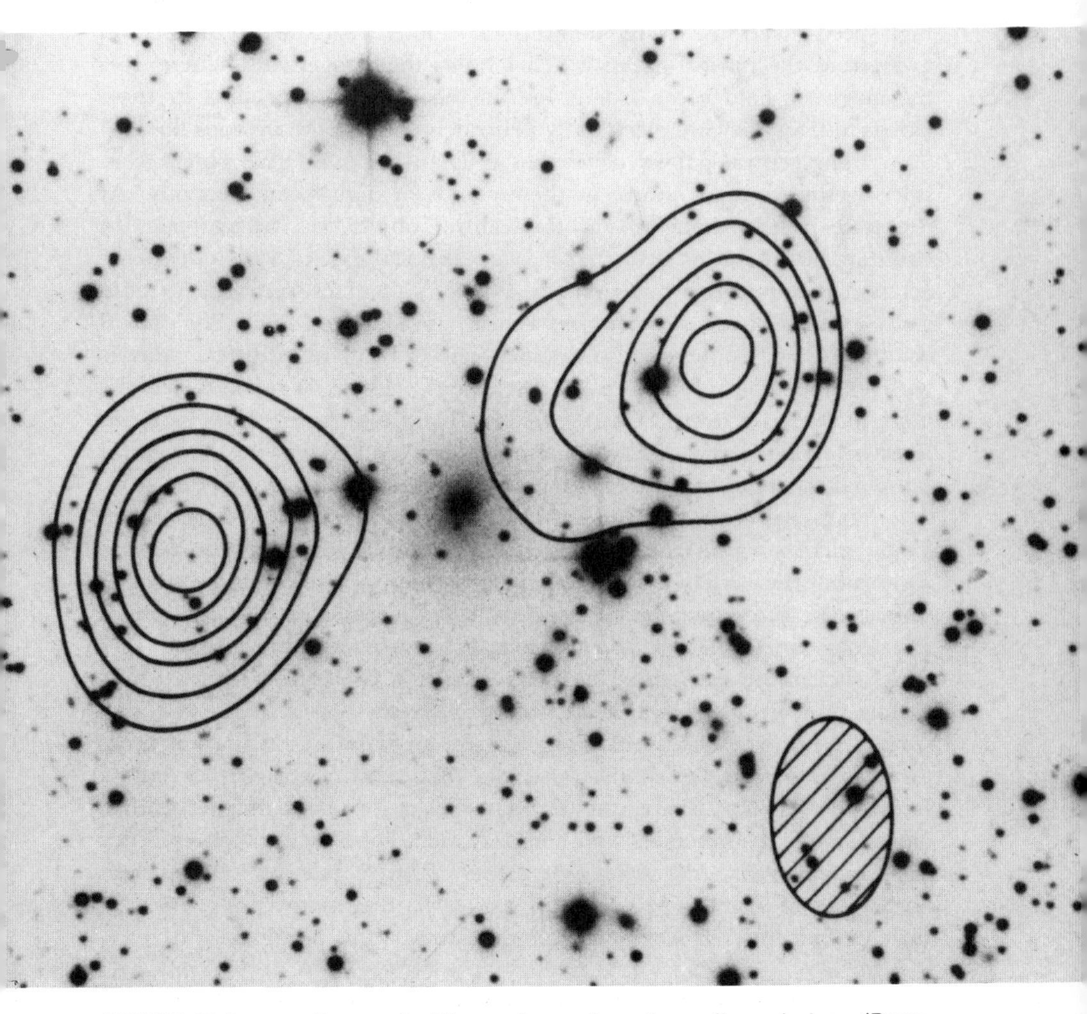

FIGURE 10-4 Cygnus A. The contours show the radio emission. (Reproduced, with permission, from "The Structure of Radio Galaxies," by A. T. Moffet, *Annual Review of Astronomy and Astrophysics,* volume 5. Copyright © 1967 by Annual Reviews Inc. All rights reserved. The photograph was taken at the Hale Observatories.)

clouds of high-energy electrons, each 17 kiloparsecs across, nearly the size of an entire galaxy. These two clouds are about 50 kiloparsecs from the central galaxy. These clouds contain a stupendous amount of energy, 10^{60} ergs, equal to the total amount of energy radiated by our galaxy in a billion years, or about one-tenth of the amount of energy released by our galaxy in its entire lifetime. Presumably these clouds were ejected by the central galaxy some 10^7 to 10^9 years ago, but attempts to model the ejection and evolution of these clouds have met many difficulties. What kind of object was able to concentrate so much energy into the form of high-speed electrons? Why don't these clouds dissipate as they travel away from the central galaxy? What holds them together? Where does the magnetic field come from? Presumably there are protons in these clouds too, for they are electrically neutral. What do the protons do?

The central galaxy of Cygnus A is an unremarkable object, considering the exciting nature of these clouds of high-speed electrons. At one time people thought that this central object was two galaxies in collision, but it is now clear that it is an elliptical galaxy with a dust lane across it that makes it appear to be double. The closest extended radio galaxy, Centaurus A, also contains a dust lane (Figure 10-5). The optical counterparts of all these radio galaxies seem to be giant elliptical galaxies with extended envelopes, called D-galaxies. These D-galaxies are the most luminous known. About half of all radio galaxies show emission lines in their spectra, indicating some form of activity in the central radio galaxy in addition to the extended lobes of radio emission.

It is not clear how the extended radio galaxies fit into the active-galaxy picture. The emitting regions are much larger than the emitting regions in other active galaxies, although some quasars contain extended sources of radio emission. Because both the extended sources, found in the strong radio galaxies, and the compact sources like the core of M 87 emit synchrotron radiation, it is very tempting to connect the two phenomena. The energy scales are vastly different, though; the extended sources are one thousand times as powerful and contain millions of times as much energy as the compact sources. It is also difficult to see how a source 10^5 parsecs in size can be related to a 1-parsec compact source. Nevertheless, the extended and compact radio sources are both evidence of galactic activity, and a complete theory of active galaxies must explain both types of source. Moving still higher in the energy scale, we now turn to the optically compact active galaxies — the N-galaxies.*

* Three developments in the radio galaxy field, reported in early 1972, strengthen the connection between radio galaxies and other types of active galaxies:

The Westerbork interferometer in the Netherlands has discovered some very large extended radio galaxies. The largest of these, 3C 236, contains two clouds of high-speed electrons, each a megaparsec long, stretching six mega-

FIGURE 10-5 NGC 5128, Centaurus A. (Hale Observatories.)

parsecs from end to end. One of these clouds could engulf the Milky Way and the Andromeda galaxy at the same time.

Very-Long-Baseline observations indicate that some extended radio galaxies contain compact electron clouds, close to the central galaxy, that are aligned with the more distant clouds. Perhaps there is some kind of connection between the two phenomena.

Observations from the Cerro Tololo Interamerican Observatory in Chile indicate that there is a very dim jet associated with the nearby radio galaxy Centaurus A (Figure 10-5). Thus there are jets in all four types of high-energy active galaxies: compact radio galaxies (M 87), extended radio galaxies (Centaurus A), N-galaxies (NGC 1275), and quasars (3C 273).

Galaxies with bright nuclei

The N-galaxies are the optical equivalents, in a sense, of the compact radio sources, having most of their luminosity contained in small, brilliant, almost stellar nuclei. A list of their properties reads very much like a list of quasar properties: spectra with broad emission lines, rapid variations, large infrared luminosity, and association with compact radio sources. There are some reasons to believe that quasars are N-galaxies that are so far away that the surrounding galaxy is invisible.

SPECTRA

Most N-galaxies, but not all, have a spectrum with very broad emission lines. A galaxy with such a spectrum is called a Seyfert galaxy. The terms *Seyfert galaxy* and *N-galaxy* are often confused. N-galaxies are distinguished by their photographic appearance, and Seyferts by their spectra. Not all N-galaxies have spectra with Seyfert characteristics.

Examine Figure 10-6, which shows some photographs of two Seyfert galaxies. In the long-exposure photos they look like normal galaxies

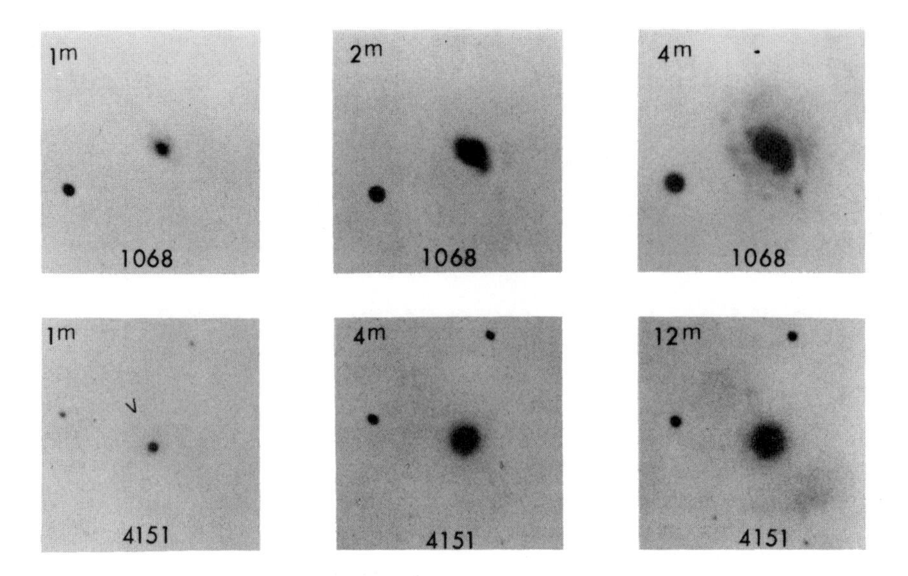

FIGURE 10-6 The Seyfert galaxies NGC 4151 and NGC 1068, with varying exposures showing that the nuclei are much brighter than the surrounding galaxy. The short-exposure photographs can pick up the bright nuclei but not the fainter surrounding galaxy. (Yerkes Observatory photograph.)

rather than N-galaxies. Yet the short-exposure photographs at the left show something that looks very much like an N-galaxy: a starlike nucleus with a little bit of fuzz around it. The nucleus of the Seyfert galaxy is much brighter than the rest of the galaxy; a one-minute exposure is sufficient to show the nucleus, while it takes a ten-minute exposure to show the underlying galaxy.

Now suppose that these two Seyfert galaxies were so far away that the only photograph we could get would be something like the left-hand panel. We should call them N-galaxies. A recent examination of this subject by William W. Morgan of Yerkes Observatory indicates that all Seyfert galaxies are N-type galaxies, in that they would be classified as N's if they were very far away. But remember, not all N's are Seyferts.

The broad emission lines in the spectra of Seyfert galaxies indicate that whatever is producing these lines is moving quite fast. As the emission-line photons leave these clouds, some are blueshifted as they come from clouds moving towards us (relative to the galaxy), and some are redshifted as they are moving away from us. (Recall the discussion in Chapter 9 about the breadth of emission lines from quasars.) The result is a smeared-out emission line, coming from clouds moving toward and away from the observer. The breadth of these lines indicates that the clouds are moving at velocities of several thousand kilometers per second. The entire emission line is redshifted by the expansion of the universe; it is the internal motions that are under examination here.

One of the puzzles, not yet solved, of Seyfert galaxies is that in some cases, not all of the lines in their spectra are smeared out. Some emission lines can be produced only in low-density regions, while others, such as the hydrogen lines, can be produced by both low- and high-density gas clouds. The lines that can be produced only in low-density regions are called forbidden lines, for the densities required are so low that it is impossible to see these lines in the spectrum of a gas in a terrestrial lab; the lab densities are just too high. The forbidden lines are often, but not always, much narrower than the hydrogen (permitted) lines. Presumably there are two different groups of gas clouds, some high-density clouds in rapid motion producing the broad hydrogen lines and some low-density clouds, moving more slowly, producing the low-density forbidden lines.

In general, the spectra of N-galaxies and Seyfert galaxies can be explained in the same way that the spectra of quasars can be. The emission lines come from gas that is recombining and ionizing as the gas is illuminated by a source of ionizing radiation. The ionizing radiation strips the electrons from the atoms in the gas cloud. When these electrons and atoms recombine, emission lines are produced as the electron cascades through the atom's energy levels. Detailed theoretical models of these gas clouds can generally explain the observed spectrum. In some Seyfert galaxies, photographs seem to show that the clouds are in the form of filaments (Figure 10-7). You can understand why this line is stronger than that line, how you can see so many different ions in the spectrum, and so on. The overall picture of the emission-line spectrum seems to be explained by the models reasonably well.

CONTINUUM RADIATION

The optical radiation from N-galaxies, like the radiation from quasars, contains both emission lines and a continuous component, not concentrated at any particular wavelengths. This continuum radiation is variable, its intensity changing over periods ranging from days to years. The separation of quasars into rapid, violent variables and quasars that vary over the years is mirrored in the N-galaxies: some vary very rapidly, and some vary much more slowly. The optical continuum from N-galaxies is polarized. In general, the continuous radiation from N-galaxies is similar to the continuum from quasars, but about ten times less intense.

Seyfert and N-galaxies are also emitting much infrared radiation. Although the total luminosity in the infrared is not known, since we have not covered the entire infrared part of the electromagnetic spectrum with observations, it seems that the infrared luminosities are extremely large. Both NGC 1068 (Figure 10-6) and NGC 1275 (Figure 10-7) put out 6×10^{46} ergs/sec of infrared radiation. Such a luminosity is more than a thousand times the optical luminosity of the entire Milky Way. Again,

FIGURE 10-7 The Seyfert galaxy NGC 1275. (Hale Observatories.)

this is a difficult area in which to make observations, but it seems clear that there is a great excess of radiation in the infrared. Its source is unknown. It is too bad that most of the luminosity of Seyfert galaxies comes out in one of the most inaccessible parts of the electromagnetic spectrum.

The N-galaxies are also the home of many of the compact radio sources mentioned in the previous section on radio galaxies. While investigation of these compact sources proceeds slowly, as Very-Long-Baseline observations take a lot of time, it is becoming clear that compact radio sources are usually associated with compact, brilliant nuclei in galaxies. Radio noise from these compact sources comes from clouds of high-energy electrons having a total energy of 10^{52} to 10^{56} ergs, much less than the total energies of 10^{60} ergs found in the quasars.

Two Seyfert galaxies, NGC 1275 and NGC 4151, are intense x-ray sources. The relation of these x-rays to other manifestations of galactic activity is quite unclear just now, because the *Uhuru* x-ray telescope is not sensitive enough to observe any but the strongest or the nearest x-ray sources. While many of the 161 sources discovered by the *Uhuru* satellite are extragalactic, only a few have been identified with active galaxies: M 87, Centaurus A, and the two Seyferts. One quasar, 3C 273 is also an x-ray source.

Unraveling the mystery of the x-rays is a task for the future. The *Uhuru* observations are quite recent, and theoreticians have not yet had a chance to interpret them thoroughly. More insight would be provided by a new, larger, more sensitive, and unfortunately more expensive satellite. These satellites, called HEAO's (High Energy Astronomical Observatory = HEAO), are scheduled to be launched in the late 1970s, but funding is uncertain in spite of the high priority given by the astronomical community. If they are launched, they will vastly increase our understanding of active galaxies, for they could see many more galaxies in more detail. The N-galaxies thus seem similar to the quasars. Are the quasars really superenergetic N-galaxies? The prevailing view among astronomers is that they are.

Links between active galaxies and quasars

Active galaxies come in different sizes, shapes, and forms. They can be ordered, as in this chapter, by the relative amount of energy given off in the form of nonstellar radiation, most of which may be synchrotron radiation from high-energy electrons in magnetic fields, and emission lines from glowing gas clouds. Figure 10-8 summarizes the different types of active galaxies discussed here.

One reason for supposing that the active galaxies depicted in

SPIRAL		MESSIER 82	
Radio	5×10^{-5}	Radio	3×10^{-4}
Infrared	0.3	Infrared	20
Optical	4	Optical	*
X-Ray	3×10^{-4}	X-Rays	?
MESSIER 87		**N-GALAXIES****	
Radio	0.09	Radio	10^{-3} to 100
Infrared	Less than 0.4	Infrared	3000
Optical	10	Optical	5
X-Ray	1.5	X-Ray	0.1 to 20
RADIO GALAXIES		**3C 120**	
Radio	0.1 to 100	Radio	0.5
Infrared	0.2 (Cen A)	Infrared	3000
Optical	10	Optical	40
X-Ray	0.03 (Cen A)		
TYPICAL QUASAR		**3C 273**	
Radio	10 to 100	Radio	300, Variable
Optical	100	Infrared	10^{5}
Infrared	***	Optical	1000
X-Ray		X-Ray	700

FIGURE 10-8 Luminosities of active galaxies. Luminosities are in units of 10^{43} ergs/sec. * = The optical luminosity of M 82 is impossible to measure because of absorption by dust in that galaxy. ** = N-galaxies are a very diverse group. *** = 3C 273 is the only quasar that has been observed in the infrared or in the x-ray regions of the spectrum. The data are from the sources listed in Note 4.

Figure 10-8 are related to each other and to the quasars is the similarity of phenomena found in many of them. In quasars we had broad emission lines, absorption lines, radio synchrotron radiation from compact clouds, a few cases of radio emission from extended regions, infrared radiation, an optical continuum that was polarized, and in one case, x-rays. Table 10-1 summarizes the extent to which these phenomena are seen in other active galaxies. It seems reasonable to suppose that the active-galaxy

TABLE 10-1

Comparison of Properties of Quasars and Active Galaxies

GALAXY TYPE	BROAD EMISSION LINES	RADIO EMISSION	INFRA-RED	OPTICAL CONTINUUM	X-RAYS
SPIRALS	X	Some	Some	X	Some
M 82	Not Broad	Some	√	?	X
M 87	Not Broad	√	√	√	√
EXTENDED RADIO GALAXIES	Sometimes	√	Cen A	X	Cen A
N-GALAXIES	√	√	√	√	NGC 1275 NGC 4151
QUASARS	√	√	√	√	√ (3C 273)

An "X" means that the phenomenon is definitely absent, at least on an interesting energy scale. A "?" means that the phenomenon has not been observed, but that observations are insufficient to decide whether a form of activity might be present on an interesting scale. A "√" means that the phenomenon is present.

phenomenon and the quasar phenomenon are related in some way. But we seek convincing evidence — some sort of a direct link between active galaxies and quasars. We can approach the search for the missing link from two directions, by looking for active galaxies that begin to resemble quasars or by looking for quasars that might be galaxies. I start with an active galaxy that almost resembles a quasar: 3C 120.

3C 120: THE MISSING LINK?

The galaxy 3C 120 is the most quasarlike of all of the N-galaxies known, so it has a place in Figure 10-8. A photograph of it is shown in Figure 10-9. The bright nucleus is about 5 kiloparsecs across, and the extended region of the galaxy is about 40 kiloparsecs in diameter, a reasonable size for a galaxy. Its nucleus is extremely luminous, putting out 3×10^{44} ergs/sec in optical radiation. The rest of the galaxy, with a luminosity of 1×10^{44} ergs/sec, has a reasonable luminosity for a galaxy. Thus 3C 120 fits the classical definition of an N-galaxy: a galaxy

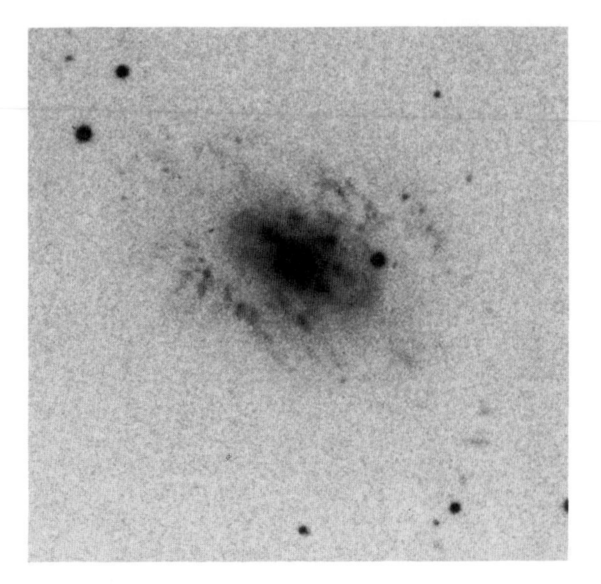

FIGURE 10-9
The N-galaxy 3C 120 photographed with the 4-meter Mayall telescope of Kitt Peak Observatory. (Hale Observatories)

having a small brilliant nucleus with a faint trace of an extended envelope. The brilliant nucleus seems to be superposed on a normal galaxy.

The spectrum of 3C 120 is quite similar to the spectrum of other N-galaxies with Seyfert characteristics. It has the very broad emission lines of hydrogen and other elements. This spectrum has been analyzed in some detail and seems to contain no real surprises as far as line intensities are concerned. It is a powerful infrared emitter, having an infrared luminosity of 10^{46} ergs/sec. This luminosity comes out at relatively short wavelengths, which can be observed from the good, dry, high-altitude infrared sites, so documentation should be reliable. 3C 120 varies quite rapidly at optical wavelengths, as it has doubled its optical brightness in a few months.

Our subject 3C 120 has been intensively investigated in the radio range by Very-Long-Baseline interferometry. It varies considerably in its radio output. What seems to be happening is that clouds of electrons are produced about once a year by the central energy source, and radio emission first appears at short wavelengths, coming from high-energy electrons. As these clouds expand, the lower-energy electrons start to radiate at longer wavelengths. The structure of the clouds is not completely determined, because VLB observers have to assume source models and then fit them to the data. At this time, it appears that 3C 120 contains a number of small components, each a few parsecs across, surrounded by a more complex structure.

Galaxy 3C 120 is one of the few active galaxies having radio variations that can be explained adequately by the simple model of expanding clouds of relativistic electrons. In most cases, while the expanding-cloud model seems qualitatively correct, attempts to match model and observation quantitatively have foundered, but here a comparison seems to work out. Remember that the expanding-cloud picture is only a simplified model. As these clouds are electrically neutral, they certainly contain protons, and the effects of the protons on the cloud evolution are unknown. Other possible cloud constituents may affect the simple model.

As 3C 120 emits about 3×10^{46} ergs/sec of nonstellar radiation, mostly in the infrared, it can be counted one of the most energetic active galaxies; it is not too much less energetic than the quasars. If you assume that the stellar radiation in the underlying galaxy is exactly equal to the luminosity in the fuzzy envelope, as shown in Figure 10-9, the stellar luminosity is 1×10^{44} ergs/sec, producing a ratio of nonstellar to stellar emission of 300. If this ratio were too much larger, or if 3C 120 were significantly farther away, this envelope would be invisible and it would look like a quasar.

This last sentence suggests another way to look for missing links between galaxies and quasars: look closely at quasars and see if there is a galaxy beneath them. One quasar has been examined with the 200-inch telescope, and this quasar, PHL 3070 (PHL = Palomar-Haro-Luyten) reveals a faint fuzzy outer envelope. It might be an N-galaxy. If so, the link between galaxies and quasars is forged. The existence of such a link would be powerful evidence that the quasar's redshift is cosmological, for it is generally agreed that galaxies are at large distances. However, the spectrum of this fuzz has not been obtained, and it is not clear that the fuzz has a stellar spectrum.

One other class of objects could provide a source for the missing link. These are the "Lacertids," named after the prototype, BL Lacertae, a variable, starlike object that shows a strongly polarized, variable optical continuum and much radio-frequency radiation. It is a powerful infrared source. These characteristics make it seem likely that BL Lacertae and the other objects like it are N-galaxies or quasars. But there is one trouble: take a spectrum of one of these, and there are no lines on your spectrum. No lines, no redshift. Without a redshift, you have no evidence that these things are even extragalactic.

Research may be near to providing a value for the distance to one of these objects. BL Lacertae is a starlike object surrounded by some fuzz. J. B. Oke and James Gunn, astronomers at the Hale Observatories, obtained a spectrum of this fuzz that looked very much like the spectrum of a galaxy with a redshift of 0.07. Such a redshift would indicate that BL Lacertae is a composite: an apparently normal elliptical galaxy with a superbrilliant starlike nucleus in the middle. The redshift indicates that this starlike nucleus is as luminous as a quasar. The only difference, then, between BL Lacertae and a quasar is that BL Lacertae does not contain any clouds of gas that produce emission lines.

Identifying BL Lacertae as a galaxy is such an important step in solving the quasar puzzle that it should be confirmed. Workers at the Lick Observatory failed to record a spectrum of the fuzzy envelope, and other observations of this object provide data that are consistent with but do not prove the hypothesis that BL Lacertae is a quasarlike object in the middle of a giant elliptical galaxy. Thus, it looks like we have an example of an extremely close connection between the quasar phenomenon and galaxies with stars in them, but we need more data to make certain.

Active galaxies show many of the same phenomena that are observed in quasars: emission lines coming from clouds of gas subject to ionizing radiation and continuum radiation from the radio to the x-ray parts of the electromagnetic spectrum. Active galaxies come in a bewildering variety of shapes and forms, and Figure 10-8 and Table 10-1 summarize their essential properties. It is very tempting to conclude on the basis of those figures that quasars are hyperactive galaxies. Most astronomers would agree with such a statement, but the proof is not quite here yet. If the observations of BL Lacertae that show that it is a galaxy with an exceedingly bright nucleus turn out to be true, an important milestone will be passed on the road connecting galaxies and quasars.

There is one important question that still has not been answered: Where does the energy come from to power these types of galactic activity? This question takes us from the real world into the model world; we have no direct observations of the central energy source, only observations of what it does. There are several models for the energy source.

APPENDIX

If you venture into the research literature, you may find some terms not mentioned in this chapter. The research literature on active galaxies is not too difficult for the nonscientist who sticks to the primarily observational papers. But the terminology may be unfamiliar.

E-Galaxy: An elliptical galaxy, here a giant elliptical galaxy with a luminosity of 10^{11} suns.

D-Galaxy: An E-galaxy with an extended halo.

Compact Galaxy: An indefinable class, embracing "galaxies which have a region of high surface brightness."[3] Because a galaxy may or may not be called compact depending on its distance from us and because many, if not most, of the galaxies discussed in this chapter are compact, it is not a useful term.

THE ENERGY SOURCES OF QUASARS

Probably the most puzzling of all quasar problems is energy. In previous chapters, enormous luminosities have been blithely tossed around. Look back to Figure 10-8, and ask what the luminosities there really mean. The amount 10^{46} ergs/sec of energy is awesome; and one quasar, 3C 273, seems to be producing a hundred times that. The luminosity of our galaxy is only about 4×10^{43} ergs/sec. The luminosities of quasars are truly stupendous — so much so that some astronomers have contended that the solution to the energy problem is to bring the quasars closer.

Another way to look at the energy problem is to consider the energy bursts in objects like 3C 120 — the purported missing link between quasars and galaxies. In 3C 120, new clouds of relativistic electrons are created once a year, and the energy in these clouds can be estimated reasonably accurately since the clouds seem to fit the simple expanding-cloud model. The energy required is 10^{52} ergs per cloud in the form of high-speed electrons. Scientific notation is sometimes deceptively innocent. The number 10^{52} does not seem so very big; why, 52 is just the number of cards in a deck, and you can hold that in one hand. Let us write it out. Then 10^{52} ergs becomes 10,000,000,000,000,000,000,000,000,000,000,000,000,000,000,000,000,000 ergs. That equals the amount of energy put out by our sun in 100 billion years, if the sun should live that long. Once every year, the central energy source in the middle of 3C 120 shoots off that much energy as high-energy electrons. It must take a remarkable object to do that. Before we consider specific models, let us briefly review quasars to see what we really know about this energy source.

General properties of the energy source

We seek some object that will produce clouds of high-speed electrons once every year or so. The size of this source is limited by the rapidity of the variations in the light of some quasars and active galaxies to a light-day or so, in these cases at least. All of the energy in the quasar comes from these high-speed electrons, and the rest of the quasar is just a conversion device. The emission-line clouds, for example, take energy from the ionizing radiation and convert it to emission-line photons. The synchrotron process taps the electrons more directly, as they spiral around magnetic field lines, converting the electron energy into photons.

There is good reason to believe that this energy source is some kind of object found at the center of galaxies. If the quasar and active galaxy phenomena are one and the same, then the existence of active galaxies makes it clear that it is in the nuclei of galaxies that we should seek the energy-producing machine. This source should be capable of producing energy in differing amounts, ranging from the comparatively weak source at the center of spiral galaxies like our Milky Way to the gargantuan source in strong quasars like 3C 273. Examinations of the evolution of quasars can provide some insight about what stage of evolution the galaxy is in when the energy source appears.

When were these energy sources most plentiful? Look out into the universe and see. You can examine various epochs of the universe by examining objects at different distances from us. Light takes time to reach the earth from a distant object. The further we probe the depths of the universe, the further we are probing into the past. We see the sun as it was when sunlight left it eight minutes ago; we see the Andromeda galaxy by light that left it two million years ago; and we see the most distant quasar known, OQ 172, as it was 12 billion years ago, as it was when the universe was approximately a billion years old (these numbers assume a Hubble constant of 50; see Chapter 14). Thus you can see how many quasars were around at various epochs by counting quasars with various redshifts, or various distances.

Maarten Schmidt of Hale Observatories undertook quasar counting. To do anything like this, you need an unbiased sample of quasars: one for which you can argue that every quasar that is brighter than some limiting magnitude has been found. Limiting your search to such a well-defined sample has the unfortunate consequence of giving you a small sample: 33 quasars, in the case of quasars producing enough radio emission to be in the 3C catalogue and brighter than optical magnitude 18.5. Counting these quasars indicates that quasars were far more abundant in the early universe than they are now. If the universe contained now the same density of quasars that it did in its first two or three billion years, several hundred quasars would be as bright as 3C 273 instead of only one.

Thus our search for an energy source is narrowed a little more — to the early stages in the nuclei of galaxies. It is not clear whether the quasars or galaxies formed first, but it is clear that quasars are very rare today. While such a constraint is not the kind of constraint we should like, a quantitative one, it at leasts puts a plausibility restriction on possible quasar models: we seek something one should not be too surprised to find in the nucleus of a young or nascent galaxy.

The energy source problem is totally unlike any other field of quasar investigation. In the previous four chapters, 7, 8, 9, and 10, the only theory involved has been model fitting. Theorists have taken observations and have been happy if a model more or less fitted the data. But here, we have very little data bearing directly on the central energy source. We can't see it. We can only see what it does, and there isn't much to go on there. In summary, we need an energy source that (1) can produce 10^{46} to 10^{48} ergs per second in the form of high-energy electrons, which burst forth in clouds every year or so, perhaps more often in some cases; (2) is small, for the variability of quasars indicates that clouds of electrons have to be a few light-days or even light-hours across in the case of the optically violent variables; and (3) is the kind of object that you would reasonably expect to find at the center of a young galaxy. Little else is known about the energy source itself. With so few constraints on the model, theoreticians are free to roam into the far reaches of the model world in their search for a possible model for this energy source. The heart of the quasar is a mysterious place.

The models presented in this chapter are described in both philosophical and chronological order. Some people want to make the energy source as prosaic as possible, while others like to let the imagination run free and see what sort of weird object it can produce. The general starting point is some kind of massive object at the center of a galaxy: a supermassive cluster of more or less ordinary stars. Some investigators believe that these stars can do the job; others envision the formation of a superstar, a Gargantua formed by coalescence of millions of stars, which will in some way belch out the necessary clouds of high-energy electrons. Others do not see the quasar phenomenon occurring until this superstar has formed a gigantic black hole, of 10^8 solar masses or so, which will produce high-energy electrons as it swallows the clouds of matter surrounding it. I finish with a very brief look at some quite radical theories: white holes, antimatter, and quarks. Please remember that these are all theories.

Colliding stars

A galaxy forms somehow in the early universe. How is not known; perhaps the quasar event triggers the formation of galaxies. Anyway, it is reasonable to expect that as the protogalactic cloud of gas collapses, a concentration of matter will form in the middle. Computer calculations of the dynamics of collapsing objects indicate that the formation of a central concentration is inevitable. To start with, let's stay fairly close to real galaxies and suppose that this central condensation is a superdense star cluster, an extreme version of what the observations of the center of our galaxy show.

It is not clear whether the stellar density at the center of this star cluster will ever reach the point for which collisions between stars occur often enough to be interesting. Princeton's Lyman Spitzer calculates that one stellar collision per year is necessary if stellar collisions are to provide enough energy (ignoring, for the moment, the troublesome issue of whether you can get relativistic electrons out of such a scheme). The density of stars has to be 10^{11} per cubic parsec for collisions to occur this often. This density is 10^5 times the density at the galactic core today. Such a star density is tremendous. Think of what it would be like to live on a planet of such a star. The night sky would be as bright as the full moon. Of course, you would run the risk of having your sun collide with another star, wiping you out of existence; and the deadly radiation coming from the heart of the quasar would be lethal to life.

It is not clear at all how a collision between two stars could produce high-energy electrons, but supernova expert Stirling Colgate has an idea. Suppose that colliding stars rapidly evolve to the supernova stage. This idea is attractive because supernovae are known to produce high-energy electrons. Maybe you don't get the high-energy electrons until *after* the supernova — thus the heart of a quasar may be millions of pulsars, ticking merrily away (Martin Rees and Jeremiah P. Ostriker, among others, take the credit for this idea). Perhaps when two stars smash into each other they blow off part of their mass as high-energy electrons. There is no shortage of ideas.

One of the reasons that these colliding-star theories are attractive is that they stay close to reality. We know something about supernovae, and they seem to be associated with the right kinds of phenomena for these theories: high-energy electrons and magnetic fields. But there is no other way to verify them. They seem reasonable, yes; but the truth or falsity of these ideas will have to wait until the future brings them into contact with reality.

Quasar superstar

When a galactic nucleus evolves to the point that stellar collisions become frequent, it is not at all clear what should happen. Would the collisions themselves cause the production of the required high-speed electrons, either by accelerating stellar evolution to the supernova stage or by some other means? Is it possible that the colliding stars coalesce and merge? It seems that the result will depend on the relative stellar velocities. If they hit each other too hard, they will blow each other up, but if the collisions are less violent, the two stars will merge.

Robert H. Sanders, working at Princeton, has followed what happens in a galactic nucleus that has colliding, coalescing stars. The colliding of small stars gradually builds up larger stars. After a few million years, one very massive star forms and grows quite rapidly; it has such a strong gravitational force that it consumes all the smaller stars that get in its way. Such a Gargantua, of 10^6 to 10^8 solar masses, is known as a superstar. While superstar evolution is not known with any certainty, it seems probable that most of the mass in the galactic nucleus would end up in such an object.

Now that the origin of this superstar has been accounted for, how is the energy explained? Different investigators have different ideas. Remember that we are trying to generate 10^{46} to 10^{48} ergs/sec in the form of clouds of high-energy electrons, thrust out at intervals.

FLARES

One way to approach the superstar theories is to suppose that superstar activity is like stellar activity, but scaled up by many orders of magnitude. Peter A. Sturrock hypothesizes that gigantic solar-flare types of eruptions will burst forth from the surface of a superstar (or galaxoid, as he calls it). We see solar flares on the sun, but they are not particularly well understood, so it is difficult to understand how to scale an unknown phenomenon to unknowable dimensions. Still, the picture is intriguing. At the surface of this vast superstar, containing 10^8 solar masses, the word goes out: *Flare!* The magnetic fields on the superstellar surface have merged, uncoupled, and merged again, but violently this time. They become entwined in each other, instabilities develop, and *Whoosh!* great streams of gas burst forth into space. (This fanciful idea is based on what happens during a solar flare; if you ever get a chance to see a movie of the sun in action, you will get some idea of what is going on.) As these streams of gas burst forth, high-energy particles accompany them. In the case of the sun, these high-energy particles move toward the earth, causing auroras and such phenomena.

Nice idea, isn't it? It seems to account for what we need. It would be better if we knew more about solar flares, however, before we become too enamored of the superflare hypothesis.

SUPER-SUPERNOVAE

The presence of high-energy particles in supernovae has caused many people to examine whether there are any possibilities in a scaled-up supernova model for the quasar powerhouse. Suppose that a superstar explodes in the same way a supernova does, producing high-energy particles. The ejected gas then forms another superstar, which explodes again. Unfortunately, it is not easy to explain some source like 3C 120, which explodes fairly often, once a year or so. Let's see if there is any more promise in a supernova remnant.

GIANT PULSARS

Probably the most thoroughly worked-out idea for the source of the high-energy electrons in quasars is the giant pulsar, or spinar, idea of Philip Morrison. Rotation will dominate many of the theories remaining for us to examine. As the superstar forms and contracts, it will spin faster and faster in the same way that a skater spins faster and faster as she pulls her arms in toward her body. Rotation can store a tremendous amount of energy in a large object like a superstar, and any mechanism that can store tremendous amounts of energy is as attractive to quasar theorists as catnip is to cats. We want energy, and rotation can store it. If you can figure out a way to get this energy out, even in small fractions, the energy-source problem can be solved.

Once again attention falls on the smaller analogues of superstars, the pulsars, and on the Crab Nebula. The Crab Nebula produces many of the same things that quasars produce, on a much smaller scale: emission lines from filaments, radio, optical continuum, x-rays, and high-energy electrons. Morrison views the quasars as giant cousins of the Crab, containing spinars, or giant pulsars. Pulsars rotate, and their rotational energy is converted into bursts of radio emission that are emitted in our direction once (or perhaps twice) in each rotation period. The mechanism for the conversion of the rotation into radio waves is far from understood, but it seems clear that large-scale electric or magnetic fields in the pulsar are somehow involved. Pulsars are also known to be the source of high-energy electrons in the Crab Nebula. All we need is magnetic fields to get synchrotron radiation, and we know that pulsars have magnetic fields, since the pulsing is electromagnetic in nature.

The spinar model has one very distinct advantage over the other models in this chapter. You have probably noticed that my descriptions of each of these ideas have ended with some comment like, "Well, it's a nice idea. But so what? How can we show that it's correct?" A spinar is rotating, and anything rotating should have some sort of periodicity. Pulsars emit radiation in pulses, so spinars should, too. Early support for the spinar idea came from the quasiperiodic variations of the quasar 3C 345. It appeared, for a while, to brighten once a year, but the regularity has not persisted. Perhaps it changed phase, entering on a new sequence. When you seek to verify the existence of a period in an astronomical object, you have to observe it over many periods – many years, in this case. More observations may settle the issue. In any event, the spinar model does have the advantage that it makes a prediction: there should be some periodicity in the variations from quasars. If this prediction can be verified, it will help the spinar idea tremendously.

The superstar models – giant flares, super-supernovae, and spinars – have the advantage of representing scaled-up versions of well-known, if not well-understood, effects. As stars are tightly packed in galactic nuclei, it seems plausible that some massive object containing between 10^6 and 10^8 solar masses, or a superstar, would form. One type of superstar theory, the spinar or giant pulsar idea, is the only quasar theory that is amenable to direct check with observations since it predicts that quasar variations should be periodic. Some theorists, however, believe that the quasar phenomenon is involved with the next phase of evolution of a superstar, its collapse towards the black hole stage.

Black hole theories

As the collapse of a very massive object proceeds, the object rotates faster and faster. No one knows exactly what happens; several poorly understood effects must be waved away as unimportant if theoretical calculations are to be practicable. There are no real supermassive objects around to allow confirmation of theory. The general consensus is that a very rapidly rotating disk would form, and that this disk could offer some promise as a quasar powerhouse. It is probable that a black hole would eventually form at the center of this disk, but it is not necessary that the hole form before the quasar phenomenon begins. It is rotation and infall that characterize this class of theories, since the material falling into and onto this disk is compressed drastically as the dimensions of the disk shrink toward the Schwarzschild radius, toward the event horizon.

The general picture is that the center of a quasar or active galaxy contains a huge disk or black hole of 10^8 solar masses or so. If the efficiency of energy generation is somewhere near the maximum theoretical efficiency, the infall of one solar mass of stuff per year would provide the necessary amount of energy. This object would lie at the center of a galaxy, eat the galaxy's core, and provide the energy for the quasar phenomenon.

Yet how can this energy come out in the form of high-speed electrons? Many theories have been proposed, but Donald Lynden-Bell's (Cambridge University, England) idea is the most intriguing and picturesque. He sees a disk of material accumulating around a central black hole in the middle of an active galaxy, swallowing material in much the same way that the black hole in Cygnus X-1 ingests the expanding envelope of its companion star. Matter accumulates in this nuclear disk. What happens next? His words take it from here:

> If too much mass accumulates in the nucleus near the central black hole the friction will try to cause the black hole to overeat. . . . [Too much mass tries to fall into the black hole. This mass gives off radiation which then pushes back on the matter now trying to fall in, pushing this new matter outward.] We suggest that this has happened in many of the most exciting cases. . . ."[1]

What you have is a black hole with a ravenous appetite in the center of a galactic nucleus, eating matter. Occasionally it eats too much and suffers from indigestion. The radiation produced by the stuff falling into the black hole reacts on the gas trying to fall in, pushing this gas out of the maw of the black hole and accelerating it to very high velocities. Lynden-Bell points out quite correctly that this scheme is too speculative to be connected to real quasars for the present. Still, it's fantastic.

RADICAL THEORIES

From time to time, some weird ideas surface in the literature. As these ideas are spectacular, they often make their way into the popular press with few caveats about their highly speculative nature. For example, white holes were first suggested in 1964 as possible quasar energy sources. White holes are time-reversed black holes (see Chapter 6). While it is theoretically possible that white holes exist, a prerequisite would be a singularity in just the right condition, ready to burst forth into the universe like a bubbling mountain spring. The idea about white holes was subsequently resurrected by the Russian astrophysicist Igor Novikov, and still later by an American, Robert M. Hjellming. Inevitably, a news article appeared in a newsweekly, claiming that we had discovered white holes. Sorry, we were only thinking about them.

Another idea for quasars is antimatter. When antimatter and matter meet, both are annihilated and all the energy turns into gamma rays. But where are these gamma rays? Gamma ray astronomers have been looking for them for years and have not found very many. Some people speculate that quarks may power quasars. Quarks are the building blocks of matter; you put quarks together and you get protons, electrons, neutrons, and all the particles that high-energy physicists have observed. Yet people have searched for quark atoms in quasars and they have not found any.

Have you had fun with this chapter? It's really enjoyable to be able to let your mind run free, not constrained by many observations, to see how you can construct a quasar powerhouse while still staying within the laws of physics. There are many, many ideas for possible quasar energy sources, and all of them work. We really don't know very much about these energy sources, so a search for them is an exploration of the model world only. There is not much sign of the real world here. It is groping in the dark; even though there are many quasar observations, these data from the real world provide little direct information about the energy source. There is no check between the model world and the real world, save for the spinar theories, and while it's fun to theorize, I feel a little uncomfortable that non-quasar specialists have occasionally taken particular models too seriously. We know so little of what is happening in the nuclei of ordinary galaxies, to say nothing of the quasars.

To briefly review the models: One group of models has a vast concentration of stars at the center of galaxies or quasars, and these stars do interesting things — collide, become supernovas, become pulsars, and so on. The superstar theories see some large object, 10^8 solar masses or so, causing the ejection of fast electrons. Maybe it flares, maybe it becomes a supernova, or maybe it is a giant pulsar (and if so, we can check on it and see whether it's periodic). A third class of models sees a disk at the center, with a black hole forming sooner or later at the center of this disk. A fourth invokes alternative energy sources, not seen elsewhere in the universe: white holes, antimatter, or quarks.

Which is correct, if any? I do not know. More than one type of phenomenon may be involved. If I were asked to bet on one of these models right now, I should bet on something that involves rotation and magnetic fields — something like the spinar model. These are all ideas. They are fun to play with, and who knows? One of them may be correct.

Because some of these models seem a bit far-fetched, it is worth considering what happens if you try to bring the quasars closer, cutting down on the luminosity required to explain their observed brightness. Current thinking along these lines raises interesting possibilities.

12 ALTERNATIVE INTERPRETATIONS OF THE REDSHIFT

One fundamental assumption underlies the interpretation of the quasar phenomenon developed in the last four chapters. The discussion has assumed that the redshifts of quasars are due to the expansion of the universe. On this assumption, astronomers base their belief that quasars are very luminous objects located at vast distances. These large luminosities can be accounted for only if the quasars are young galaxies with powerful explosions taking place at their centers. This fundamental assumption is now being tested as some physicists and astronomers argue that the redshifts are due to some other cause than the expansion of the universe, some cause now unknown to physics. In the challengers' view, quasars are somewhat smaller, less luminous objects that are the scene of this strange new phenomenon that produces the redshifts. If the challenge is sustained, we shall have discovered a new physical law.

The status of the challenge to the traditional view that the redshifts are cosmological is not clear at present. While most people still support the traditional view, the evidence in favor of the challenge is sufficiently strong that it cannot be dismissed lightly. As the weight of the evidence is insufficient (according to the majority view) to disprove the cosmological viewpoint, the challenge has not yet succeeded. Its outcome is unsettled. Whatever happens nevertheless, the challenge will have helped quasar research. The cosmological redshift must now be proved rather than accepted on faith, and the search for evidence supporting the cosmological redshift has illuminated the relation between quasars and galaxies.

When quasars were first discovered, many theorists explored various alternative causes of the redshift. One idea was that the redshifts are gravitational; this idea fails, for all of the spectrum lines from the quasar have the same redshift and the emitting gas cloud must be unacceptably small if the gravitational redshifts are to be the same at all points in the cloud. Another idea was that quasars were objects ejected at high velocities from our galaxy. We should expect that in such a case other galaxies would eject quasars, some of which would be traveling towards us and show a blueshift. No such objects have been observed. The failure of these two attempts to find an alternative cause for the redshift points to one inescapable conclusion: If the quasar redshifts are due to some physical law now known, they are cosmological.

The challenge

The cosmological idea is not without its problems, either. They first became evident in 1966, when Fred Hoyle of Cambridge University and Geoffrey Burbidge and Wallace L. W. Sargent, then of the University of California, pointed out the problem that the rapid variability of some quasars raised. (This problem is discussed in Chapter 9.) Briefly, the rapid variations of the brightness of some quasars indicates that they are very small. It is difficult to understand how the high luminosity in quasars can come from such a small object. The problems that were pointed out in 1966 can be avoided with more complex models, as was shown subsequently. But some doubt had been cast on the cosmological-redshift theory; two of the authors of the paper, Hoyle and Burbidge, have since been prominent in the challenge to the traditional explanation of quasar redshifts.

But how can the challengers prove that quasars are closer than the redshifts indicate? The centerpiece of their evidence is a group of photographs taken by H. C. Arp of the Hale Observatories. Arp's photographs seem to show connections between several quasars and galaxies with much lower redshifts. These connections, if real, imply that the quasar and the galaxy in each case are at the same distance from the earth. As the redshift of the galaxy is much smaller than the redshift of the quasar, the inevitable conclusion is that some of the quasar's redshift is "extra," or to use the current term, discordant, and the cosmological-redshift theory is thereby demolished.

Yet two objects close together in the sky may or may not be real companions, as we can see the universe only from one vantage point — the earth. For example, Mizar, the second star from the end of the handle of the Big Dipper, appears to have two companions (Figure 12-1). One of these companions, Alcor, is distant, whereas Mizar B is so close to Mizar A that you need a telescope to split them. When the system is analyzed, it turns out that only Mizar A and B are true companions, while Alcor is a background star, much farther away.

Thus if we seek to show that quasars are associated with lower-redshift galaxies, we must show a real physical connection between the objects. One connection has been investigated more thoroughly than others, and I present it as an example.

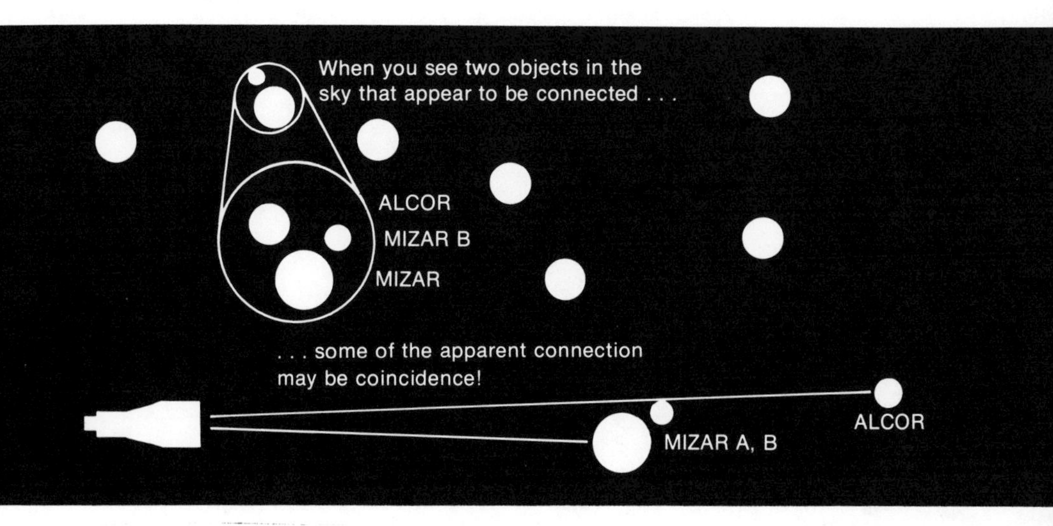

FIGURE 12-1 Two stars, close together in the sky, may be far apart in space, as Alcor and Mizar in the Big Dipper are.

MARKARIAN 205 AND NGC 4319

The remarkable pair of objects Markarian 205 and NGC 4319 is shown in Figures 12-2 and 12-3. Galaxy NGC 4319 is the large spiral galaxy at the top, and it has a redshift of 1800 kilometers per second, or $z = 0.006$. Markarian 205 is the starlike object below the galaxy, and it has a quasar spectrum with a redshift of 21,000 kilometers per second, or $z = 0.07$. If these two objects are true companions, most of the redshift of Markarian 205 is extra, or discordant, coming from some physical cause now unknown to science.

But are these two objects true companions? Two possible views are shown in Figure 12-4. How are we to decide? Look at Figures 12-2 and 12-3 again, the photographs of the objects, and concentrate your attention on the area between the galaxy and the quasar. There seems to be a bridge of luminous material connecting the quasar and the galaxy. These photographs can be most straightforwardly interpreted by supposing that this luminous matter connects the galaxy to the quasar. Such an interpretation indicates that the quasar and the galaxy are at the same distance, and that the quasar has a discordant redshift.

Yet many of us hesitate to call for new laws of physics until all attempts to explain the observations by the old laws have failed. Other explanations of the observations have appeared:

1. Maybe it isn't there after all. Some investigators have failed to photograph this bridge when they have photographed these objects.

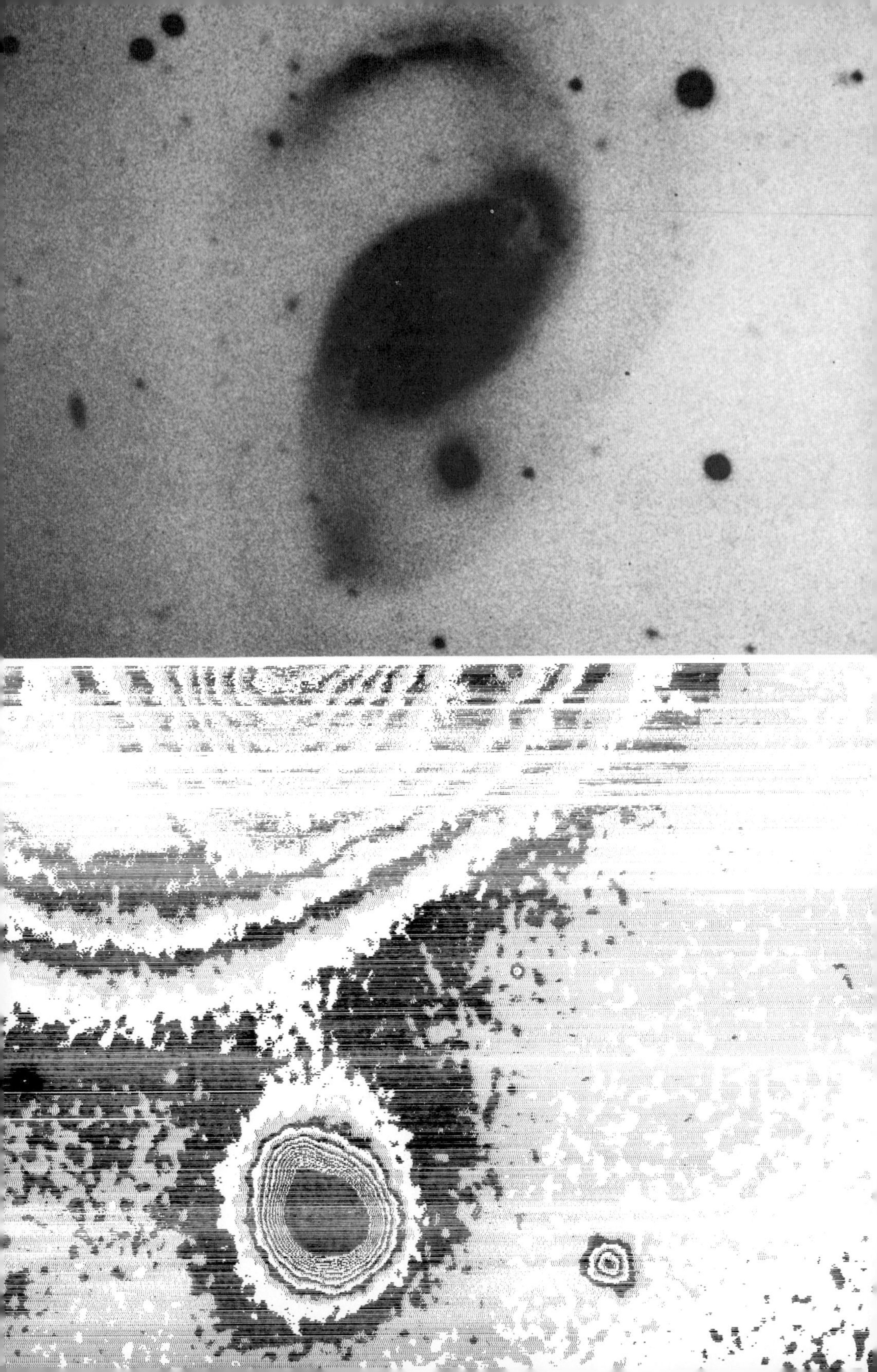

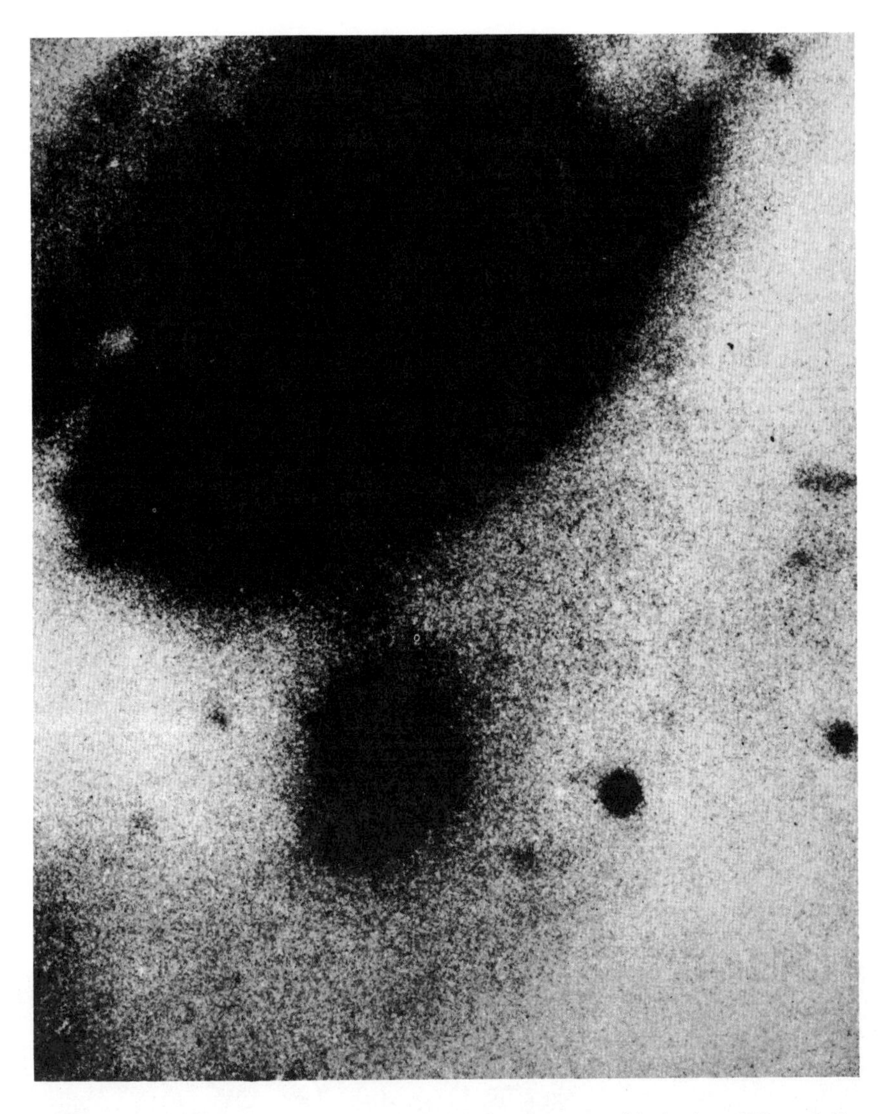

FIGURE 12-3 A closer look at the relation between Markarian 205 and the galaxy NGC 4319. (From C. R. Lynds and A. G. Millikan, *Astrophysical Journal Letters,* vol. 176, page L5, 1972, published by the University of Chicago Press.

FIGURE 12-2 Markarian 205, a quasar, and NGC 4139, a spiral galaxy. (From C. R. Lynds and A. G. Millikan, *Astrophysical Journal Letters,* vol. 176, page L5, 1972, published by the University of Chicago Press.

2. Perhaps it is just the overlap of the images of the galaxy and the quasar, since images can blend together in the photographic process.

3. It is a short, stubby spiral arm of NGC 4319 that just happens to point right at the quasar.

4. It is a background galaxy that happens to be there, fooling us by looking like a connection between the galaxy and the quasar.

What then is the answer? If you are like most of my students, you are probably feeling a little uncomfortable. They request, "Tell us the real story." I cannot give you a definite statement one way or the other because I feel unsatisfied too. I have looked and looked at these pictures and read and reread the papers in which they appear in the hope of giving you something more definite than the academic hedging of the last few pages. Most astronomers, myself included, feel that there are several ways of explaining the evidence without invoking discordant redshifts, so we are unwilling to question the cosmological interpretation of the redshifts on the basis of one case alone. But there is more evidence. Two more galaxy-quasar associations exist, but (in my view at least) neither is as compelling as the Markarian 205–NGC 4319 association. There are also a few cases in which the redshifts of galaxies seem discordant, and I now turn to one of those.

STEPHAN'S QUINTET OR QUARTET

Figure 12-5 is a photograph of the remarkable little group of galaxies, named after its discoverer, M. E. Stephan of Marseilles. Figure 12-6 shows the most unusual property of this little group, the redshifts of

FIGURE 12-4 Is Markarian 205 really a companion of NGC 4319?

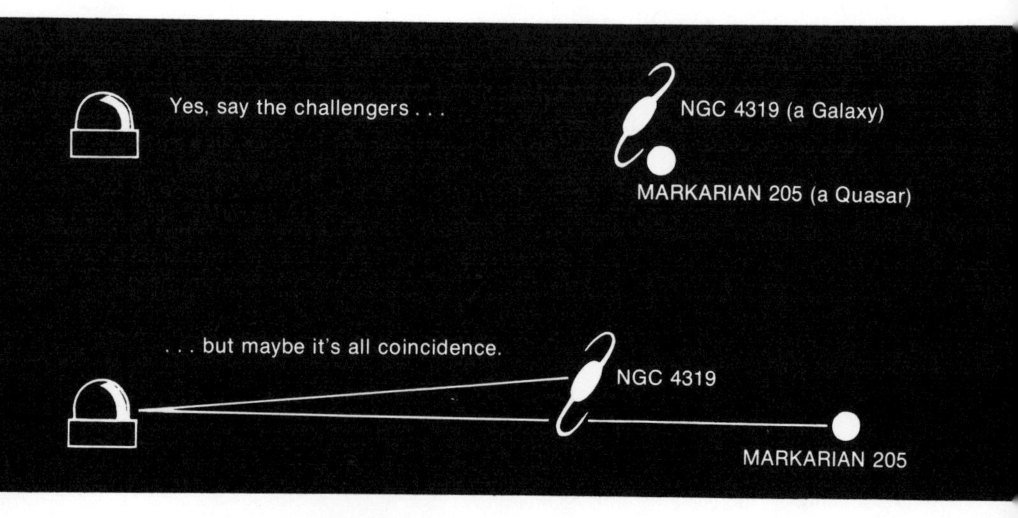

FIGURE 12-5 Some call this group of galaxies Stephan's quartet and some call it a quintet. The galaxy at the lower left may be a foreground galaxy. (Hale Observatories photograph.)

the different galaxies. It is quite surprising that the redshift of NGC 7320 is so much smaller than the redshift of any of the others. Challengers of the cosmological-redshift picture point to this group as another example of a group of galaxies, apparently companions in space, for which the redshift of the galaxies does not seem to be a valid indicator of distance.

Yet once again, it is possible to visualize the situation somewhat differently. Maybe the group should really be called Stephan's Quartet, since only the four high-redshift members form a true physical grouping. In this interpretation, NGC 7320 is a foreground galaxy, much closer to us than the quartet. As the conventional view holds that a high redshift means a large distance, the quartet interpretation is consistent with it, for the high-redshift quartet is far away and the low-redshift NGC 7320 is closer.

Because we think we understand galaxies better than we understand quasars, it seems that we could try to appeal to some other way of measuring the distances of these galaxies. Such attempts have been made,

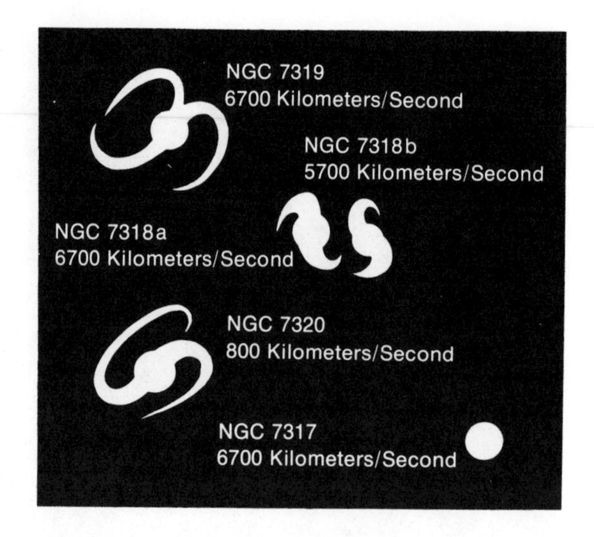

FIGURE 12-6
Redshifts and names of the galaxies in Stephan's Quartet/Quintet.

but again the results are not entirely well defined. There is some evidence in favor of the quartet (conventional) interpretation and there is some evidence on the other side. Perhaps the fairest assessment of the situation is to say that there is more work to be done on these galaxies, but that the conventional view has enough evidence to support it for the power of the challengers' arguments to be blunted.

Statistics: big ideas from small numbers[1]

The heart of the disputes about peculiar objects like Stephan's Quintet/Quartet and the Markarian 205–NGC 4319 pair is the question, Are these two objects really associated with each other or is the line-up just due to chance? You might think that the tools of statistics would be good to use to attack this question, since statistics do not involve the judgment of individual investigators. Statistical equations have no axe to grind. Yet the investigators who use the equations do have an axe to grind, so that statistical arguments have not proved of much assistance in settling the controversy. There are two fundamental weaknesses in applying statistical arguments to quasars; you are not trying to test hypotheses that were formulated before the data appeared, and you are dealing with small numbers of objects.

You can use statistics to answer questions only if the question has been precisely formulated before you go out and seek the data. It is erroneous to say, as some have done, that there is one chance in a thousand that Stephan's Quintet is a chance aggregation of galaxies. If you make such a statement, you are reasoning in circles. You take a photograph (or measure redshifts), notice a strange phenomenon, and then use statistics to prove that you have found a strange phenomenon. You must use your judgment to decide whether this strange phenomenon is meaningful; statistics are no help.

Statistics can also be used in a different way. Take a list of quasars and ask, "Are these quasars closer to bright galaxies than you would expect from chance alone?" This is a legitimate approach, for you have formulated the question before you look at the data. You can now use statistics to try to find an answer.

Yet once again you encounter problems. Many investigators have used the above approach, and they produce contradictory results. Whether a statistical investigation supports the challenge or not depends on the list of quasars used. The reason that statistics are ambiguous here is that the number of quasar-galaxy associations is so small. For example, one heralded investigation showed that 5 out of 50 quasars in a sample chosen from the 3C catalogue were quite close to bright galaxies, a result that had a chance probability of 1 in 100,000.[2] Yet five quasars is a very small number on which to base a conclusion. Other investigations have failed to find any statistical correlations.[3]

Where does this leave us? You have to look at each case and use your judgment. Most astronomers do not believe that the case for noncosmological or discordant redshifts is strong enough at this time to be used as the basis for some future work. I agree with this view, and while I have tried to present the arguments in favor of discordant redshifts fairly, I recommend Halton C. Arp's article in *Science* as an eloquent and readable expression of the case for noncosmological redshifts.[4] Arp's article is written at the same general level as this book.

Support for cosmological redshifts

The challenge to the cosmological redshift picture will have contributed a great deal to our understanding of galaxies and quasars even if it turns out to be wrong. For supporters of the traditional view have not been content to find weaknesses in the arguments in favor of the discordant-redshift theory; they have sought and found independent evidence that quasars are at the distances that their redshifts indicate.

So far, three quasars have been found in clusters of galaxies where the redshift of the galaxies is the same as the redshift of the quasar. As the galaxy redshifts are generally accepted to be valid indicators of distance, the quasar is as far away as its redshift indicates. Figures 12-7 and

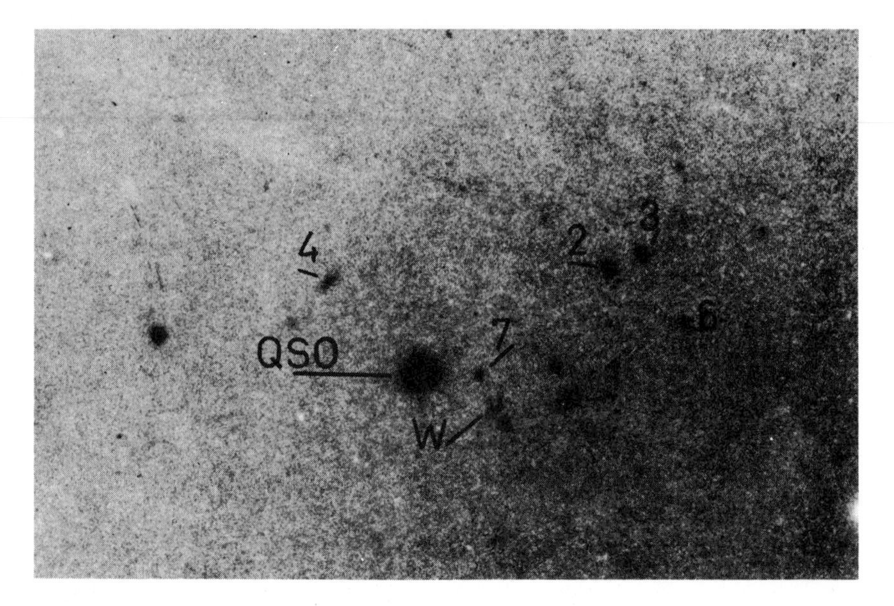

FIGURE 12-7 The quasar PKS 2251+11 surrounded by a cluster of galaxies. The galaxies are numbered, and the quasar is labeled "QSO." (From J. E. Gunn, *Astrophysical Journal Letters,* vol. 164, p. L113, 1971, published by the University of Chicago Press. Copyright © by the University of Chicago. All rights reserved.)

12-8 show two of these quasars, PKS 2251 + 11 and 3C 323.1. Supporters of the discordant-redshift view have advanced counterarguments – maybe the associations are coincidental, maybe the fuzzy images around PKS 2251 + 11 are spurious, and so on. Yet these arguments seem contrived. The third association involves the quasar RN 8.

The discovery of BL Lacertae also supports the cosmological viewpoint. (BL Lac was the object that had no emission lines in its spectrum and turned out to be a quasarlike object in the middle of an ordinary galaxy.) Here is an object that has many of the features of a quasar – high luminosity, rapid variability, and high-speed electrons – intimately associated with a galaxy. The discovery of this object supports the theory that quasars are violently exploding galaxies.

You can try to settle the argument by agreeing with both sides, theorizing that there are two types of quasars: one with cosmological redshifts and one with noncosmological redshifts. Such ideas have occasionally cropped up in the literature, but both participants in a recent scientific debate on the controversy agreed that the coincidence would be too much to accept. At this time, the idea that there are two fundamentally different types of objects with similar properties as seen from the earth seems to most people to go beyond the evidence.

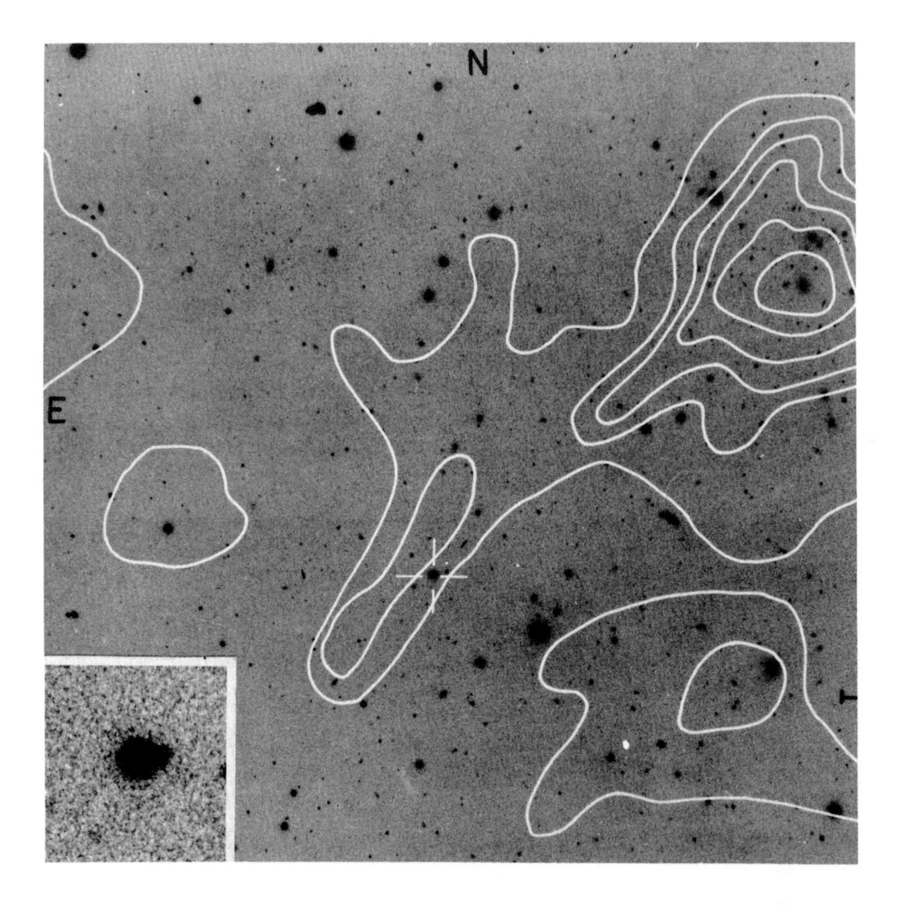

FIGURE 12-8 The quasar 3C 323.1 at the edge of a cluster of galaxies. The cluster of galaxies is delineated by contours. *Inset:* A closeup of the quasar. *(From* A. Oemler, Jr., J. E. Gunn, and J. B. Oke, *Astrophysical Journal Letters,* vol. 176, p. L47–L51, 1972, published by the University of Chicago Press. Copyright © by the American Astronomical Society. All rights reserved.)

Discordant redshifts and scientific revolutions

The challenge to the accepted idea that the redshifts of all extragalactic objects are cosmological and the astronomical community's response to that challenge fit the pattern of scientific revolutions described by Kuhn and paraphrased in the first chapter. That is, a scientific paradigm, or accepted viewpoint, is always beset by anomalies in which the model based on the paradigm does not quite match the observations. When normal science fails to bring model and observation closer together by refining the model or reexamining the observations, a crisis results, in which people question the underpinnings of the model. If the model and the observational results turn out to be irreconcilable, scientists who are not too attached to the original paradigm (generally a somewhat irreverent group) search for new scientific laws, which may form the basis for a new paradigm if models based on the new paradigm can explain the anomaly and conform with existing observations.

NORMAL SCIENCE

The pattern of normal science Kuhn described corresponds well with the way that the challenge was posed and with the way that science responded. Once the paradigm of cosmological redshifts was established, most astronomers expected that the redshifts of all quasars would be cosmological, in accordance with the paradigm, or pattern. The first articles containing evidence for noncosmological, discordant redshifts were not accepted by the most prestigious journals because astronomers demanded a higher standard of proof for such paradigm-shattering ideas. Yet the anomaly persisted.

CRISIS

The anomaly of discordant redshifts is not quite at the crisis stage, where serious questioning of the current paradigm begins. The challengers are still trying to prove their case; people have not seriously tried to construct new physical laws to explain the discordant-redshift phenomenon. While a few speculations have appeared, most people on both sides of the controversy are still preoccupied with establishing the reality or unreality of the discordant-redshift phenomenon. We are still in an era of normal science, where we test and retest the existence of discordant redshifts. The anomaly is still there, but it is not yet time to enter a crisis atmosphere and search for new paradigms to replace the cosmological-redshift paradigm.

Kuhn's model of a scientific revolution imparts a certain timeliness to different phases of the scientific process. Since science is a pluralistic enterprise, any individual scientist can do anything he feels like, but certain types of research will have more impact on the scientific community than other projects. Alternatives to the current paradigm make little impact on the progress of science unless a crisis exists, and we are not there yet in the discordant-redshift story. The anomaly is still being explored as the controversy has been focused on one question: Is the cosmological redshift scheme correct, or do discordant redshifts exist? The issue is confined as Kuhn's scheme says it should be.

Believers in discordant redshifts have challenged the accepted view that all redshifts are due to the expansion of the universe. The issues in the debate are summarized on page 216. The challenge is fascinating, as success in the challenge would mean that some new physical laws would have to be invented. Fred Hoyle explores the landscapes opened up by the challenge:

> The implication is that cosmology will have a great deal to say about fundamental physics. This is a conclusion which many astronomers and physicists are trying to resist. . . . For each of us there is a decision to be made. Do we cross a bridge into wholly unfamiliar territory or do we try to remain safely within well-known concepts? This depends on how each of us sees the data.[5]

As I have worked on this chapter, I have seen this fascinating bridge in front of me. But does it lead to the Promised Land or off a cliff? Too many times in the past such tantalizing bridges have appeared, only to fade into obscurity as the observations turned out to be misinterpretations of the data. Fred Hoyle sees the situation one way: "There seems to me no doubt that the trend is forcing us, whether we like it or not, across this exceedingly important bridge. Either the bridge must be crossed or one must judge the data of the past five years to be extremely freakish."[6]

Yet there is another way to judge the data of the past few years, expressed by John Bahcall: "Seek and ye shall find, but beware of what you find if you have to work very hard to see something you wanted to find."[7]

Evidence Bearing on the Redshift Controversy

EVIDENCE BOTH SIDES CAN EXPLAIN

Continuity between quasars and active galaxies (Chapter 10)

> *Conventional:* If all redshifts are cosmological, then the sequence of Figure 10-8 is one of increasing energy.
>
> *Discordant:* If the quasar redshifts are noncosmological, then quasars are more compact, possibly less luminous analogues of the active galaxies.

Problems in understanding what is going on inside quasars: high luminosity of small objects, the nature of the energy source, and so on (Chapters 8 and 9)

> *Conventional:* One can work around the problems by using suitably complex models. Anyway, the problems that still exist refer to our poor understanding of the quasar phenomenon, not a poor understanding of the quasar distance.
>
> *Discordant:* Problems are relieved if the quasars are closer.

EVIDENCE FAVORING COSMOLOGICAL REDSHIFTS

Galaxies underlie quasars; close inspection of quasar images shows that there is fuzz, presumably a galaxy, surrounding them

> *Rebuttal:* How do you know the fuzz is a galaxy? It must have the same redshift as the quasar.
>
> *Answer:* BL Lac does! (If the BL Lac evidence holds up, it is a very strong piece of evidence in favor of cosmological redshifts.)

Several quasars are found in or near clusters of galaxies with similar redshifts (PKS 2251 + 11, 3C 323.1, RN 8—Figures 12-7 and 12-8)

> *Rebuttal:* The associations may be due to chance, as the quasars are on the edge of the clusters of galaxies.

Where do the redshifts come from if they are not cosmological?

> *Rebuttal:* "So we have discovered something new."[8]
>
> *Answer:* Yes, but the burden of proof is greater for those who wish to discover new laws of physics.

EVIDENCE FAVORING DISCORDANT REDSHIFTS

Quasar redshifts tend to come in multiples of 0.031

> *Rebuttal:* There is a 2.5 percent probability that this is due to chance.

Statistical correlations exist between quasars and bright galaxies

> *Rebuttal:* The statistical results you obtain depend on the particular list of quasars and bright galaxies that you choose.

Luminous bridges between quasars and bright galaxies (for example, Markarian 205 — Figures 12-2, 12-3, 12-4)

> *Rebuttal:* The bridges may not be there. They may just be the result of overlapping images.

Discordant redshifts in groups of galaxies (for example, Stephan's Quartet/Quintet — Figures 12-5, 12-6)

> *Rebuttal:* This group is really Stephan's Quartet with a galaxy superposed in the foreground.

SUMMARY OF PART TWO

The completion of the 200-inch Hale reflector at the end of World War II marked the beginning of the intensive study of extragalactic objects that has paid great dividends for twentieth-century astronomy. Quasars, the most luminous objects in the universe, were discovered in 1963. Since then, we have realized that there are numerous objects similar to the quasars: the active galaxies. Many mysteries remain in our work on these objects.

The following display summarizes the essential results of this section, again differentiating among fact, concrete theory, informed opinion, and speculation. Here the observers are in the forefront of research, so that most theory consists of models that are very closely tied to the real world, in marked contrast to the black hole business where we are not even sure black holes exist.

One basic question remains: Are the quasars really at cosmological distances? Most astronomers believe so, but such a fundamental question as this needs a solidly proved answer. The future may bring one.

OBSERVATIONAL FACT	Quasars and active galaxies exist, with emission lines, radio-frequency radiation, infrared excesses, optical variations, and x-rays
	Very-Long-Baseline observations of the radio structure of these objects show both compact and extended sources
	There is activity at the center of the Milky Way galaxy
CONCRETE THEORY	Radio-frequency radiation from quasars and active galaxies is synchrotron radiation
	The emission-line spectra of these objects come from clouds of gas subject to ultraviolet, ionizing radiation

INFORMED OPINION	Variations in the intensity of quasar continuum radiation come from the production and expansion of clouds of relativistic electrons (fairly well confirmed for radio radiation, not so for other radiation) The redshifts of quasars are cosmological (agreed to by most astronomers but not all)
SPECULATION	What is the source of the infrared radiation? What is the nature of the central energy source?

THE UNIVERSE

Probably all astronomers, from the first caveman to follow the days, nights, and phases of the moon by making notches in a bone fragment to the twentieth-century astronomer using the latest gadgetry on the 200-inch telescope, became astronomers to determine man's place in the universe. Yet to understand the universe, you have to understand galaxies, and to understand galaxies, you have to understand stars, somewhat at least. Furthermore, the evolution of the universe is governed by gravity, so an understanding of the way gravity works is prerequisite to accepting the challenge of cosmology, the study of the evolution of the universe.

Until the twentieth century, cosmology was largely the province of philosophers. People continued to learn more about the nearby planets and stars, but really knew nothing about the vast expanses of the universe. Observational cosmology is a product of the last seventy years; we have finally developed the tools to pick up some observational facts bearing on the evolution of the entire universe. Cosmology is no longer limited to the model world — real data can be brought into the picture and various models can be confirmed or disproved.

Here, as in Part One (Black Holes), the story begins with the model world, with Chapter 13 describing the cosmological model that almost all astronomers believe in these days—the Big Bang theory. Subsequent chapters show why this theory enjoys such widespread support, as the model world is compared with the real world. Chapter 14 examines the age of the universe, the time interval that has passed since the Big Bang. In Chapter 15, the observational facts that confirm the Big Bang picture are examined, and the evidence bearing on the Big Bang theory's one-time rival, the Steady State theory, is discussed. Chapter 16 looks to the future: Will the expanding universe expand forever, or will it slow down, stop, and start contracting it? At this time, we do not know what it will do.

LIFE CYCLE OF THE UNIVERSE: A MODEL

To study the life cycle of the universe, we begin with theory, or a model history of the cosmos. The universe is expanding now and has been expanding for a long time. Any hypothetical history of the universe must account for this expansion, and also for the observed fact that the universe looks generally the same in every direction. If you follow the expansion backward in time, you see that the universe must have been smaller in the past. Following it back still further, you find that there was a time when all the matter in the universe was packed tightly together. The explosion of this cosmic egg provided impetus for the expansion that we now observe. This theory is the Big Bang theory, which is currently accepted by all but a handful of astronomers. (Chapter 15 contains a brief discussion of a one-time rival, the Steady State theory and a discussion of the evidence against this theory.) Between the Big Bang explosion, some 10 to 20 billion years ago, and the present, some interesting events have occurred, and these occurrences have had observable consequences that allow us to verify the model.

The evolving universe

The discovery, made in the 1920s, that the universe was expanding provided the foundation for modern cosmology. Follow the expansion backward, and you come to the Big Bang, the explosion that marked the beginning of the expansion. Follow it forward, and maybe the expansion slows to a stop and maybe it goes on forever.

The idea of an expanding, or evolving, universe is new in Western thought. Most European philosophers have generally believed that the universe is static, unchanging. "One generation goeth, and another generation cometh, but the earth abideth forever. . . . That which has been is that which shall be, and that which has been done is that which shall be done; and there is no new thing under the sun" (Eccles. 1:4, 9). Others through history have expressed this same thought, but none so eloquently as the preacher. The discoveries of the 1920s overturned this idea; they showed that the universe does indeed evolve as it expands. The Steady State theory, now in conflict with the evidence, was an attempt to create an expanding but unchanging universe.

The evolution of the expanding-universe theory illustrates the interplay between the model world of the theorists and the real world of the observers. In 1917, two years after the presentation of the General Theory of Relativity, Albert Einstein and the Dutch physicist Willem de Sitter independently applied the new theory to the universe as a whole. Their research showed that a universe that obeyed Einstein's theory of gravity must either expand or contract; it could not remain static. Because of the Western predilection for an unchanging universe, this idea was quite alien. Einstein even tried to modify his theory so that the universe could remain static, making what he later called "the greatest blunder of my life."[1] The Russian mathematician Alexander Friedmann ignored Einstein's modification and calculated some model universes as a mathematical exercise. These models have endured; at the current level of sophistication, they are the best models of the universe that we have.

The observers were not so far behind. In the early 1920s, American astronomers in California were making considerable progress in extending the frontiers of observational astronomy beyond the Milky Way. It became clear that the wispy spiral nebulae were galaxies, each one containing billions of stars. Looking at one of these galaxies can give you some perspective on the vastness of the universe, as you realize that this fuzzy little patch is a galaxy equal to our Milky Way. If you can, go somewhere out in the countryside on a dark autumn night and try to find the Andromeda galaxy, the nearest large spiral. (Directions for finding it are given in Figure 13-1.) You do need to be some distance away from the glare of city lights to see it. As you look at this little fuzzy patch of light, think how many stars you are seeing: 100,000,000,000 of them, each one somewhat like the sun. Binoculars will allow you to see the galaxy a little better.

The evolution of the real universe was revealed later in the 1920s, when Edwin P. Hubble of the Mount Wilson Observatory extended earlier measurements, by Vesto M. Slipher, of the spectra of distant galaxies and found that all but a few of the nearest galaxies were speeding away from the earth. Hubble and his colleague Milton Humason discovered the law that bears Hubble's name: Velocity of recession equals Hubble's constant times the distance of the galaxy. If Einstein had not modified his original theory in 1917, he would have made one of the most stupendous predictions in the history of science: the prediction of the expanding universe. As it was, the expanding universe was discovered in due course, so it is unclear whether the history of science would have been substantially affected if Einstein had not erred.

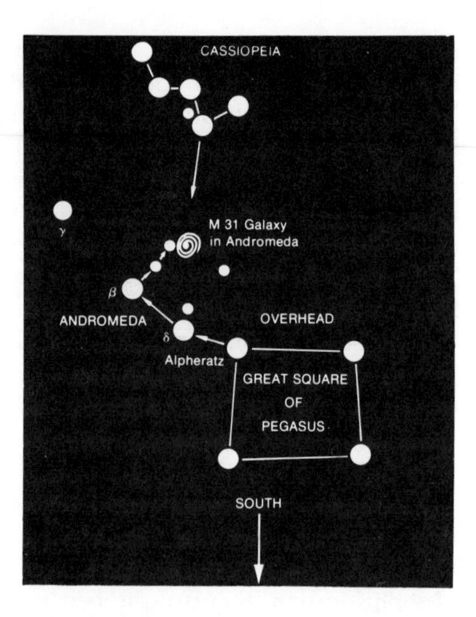

FIGURE 13-1 How to find the Andromeda galaxy. At about 9 p.m. in late October, look high in the southern sky, almost overhead, and find the Great Square of Pegasus, marked by four moderately bright stars and about 15 degrees on a side. Messier 31, the Andromeda galaxy, is northeast of the Square, towards the W- (or M-) shaped constellation of Cassiopeia. From the northeast corner of the Square, find δ (delta) and β (beta) Andromedae; let your eyes make a right turn at β and go towards Cassiopeia. You will find a fuzzy star, the Andromeda galaxy, which you can see if the sky is dark.

THE NATURE OF THE EXPANDING UNIVERSE

Two questions about the nature of the expanding universe immediately arise. How does one explain Hubble's law, that the expansion rate *increases* as one inspects areas farther and farther from the earth? Since the expansion is uniform in all directions, aren't we at the center of the expansion? These two questions are naturally answered if you realize that it is the entire universe that is expanding. You have to think about it for a little while.

Writers on cosmology have concocted different analogies to make the expansion of the universe seem more concrete; I shall mention two of them and hope that you will find that at least one helps you understand how each galaxy, in an expanding universe, sees all the others recede, following Hubble's law. In Figure 13-2, the universe is seen to be analogous to a raisin cake, with the raisins representing the galaxies. As the universe expands, or as the cake rises in the pan, the galaxies (or raisins) get farther apart. The analogy that seemed most appropriate to

When a raisin cake rises in a pan, each raisin moves away from its neighbors as the dough expands . . .

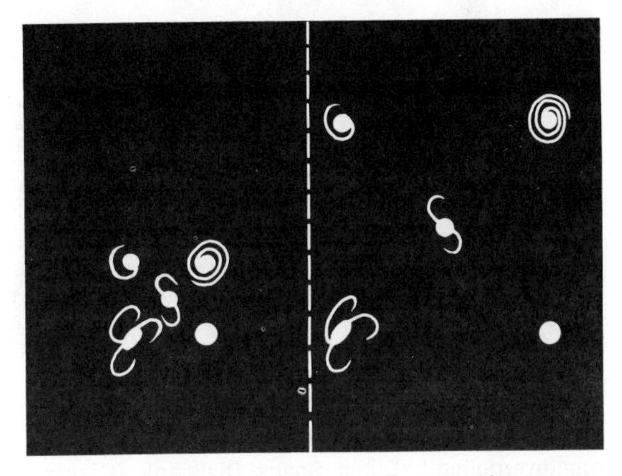

. . . just as, in the expanding universe, every galaxy moves away from its neighbor galaxies.

FIGURE 13-2
The expanding universe, seen analogous to a raisin cake.

me, when I first saw it, is George Gamow's jungle-gym analogy (Figure 13-3). Gamow likened the universe to a gigantic jungle gym made of telescoping pipe. The galaxies are represented by children (or in the figure, Escher's cubes) sitting at intersections of the jungle gym. As the pipe expands, the whole framework expands, moving each cube away from all the others, since there is more pipe between it and each of its neighbors as time goes on. Every cube seems to be in the center of the expansion, but there really is no center as the entire framework is expanding. The idea that it is the framework that is expanding is the key to the concept of the expanding universe.

A little more thought with one of these analogies will produce the Hubble expansion law: that the rate of recession increases with distance

FIGURE 13-3 The analogy between the expanding universe and an expanding three-dimensional framework is well illustrated by M. C. Escher's "Kubische ruimteverdeling" (Cubic Space Division). (Courtesy of the Escher Foundation—Haags Gemeentemuseum—The Hague.)

(see Figure 13-4). Here we follow the expansion over the time that it takes the universe to double in size. The farther apart the two galaxies at the beginning of the picture, the more space between them to expand and the more the change in separation. Since the velocity is just a measure of the rate at which the two galaxies separate, the velocity is higher for the galaxies that were more distant from each other at the beginning. Figure 13-4 contains some numerical examples.

But where did it all begin? A Belgian cleric, the Abbé Georges Lemaitre, made a suggestion in the 1920s which has subsequently been accepted by most astronomers, as it has been shown to agree with observa-

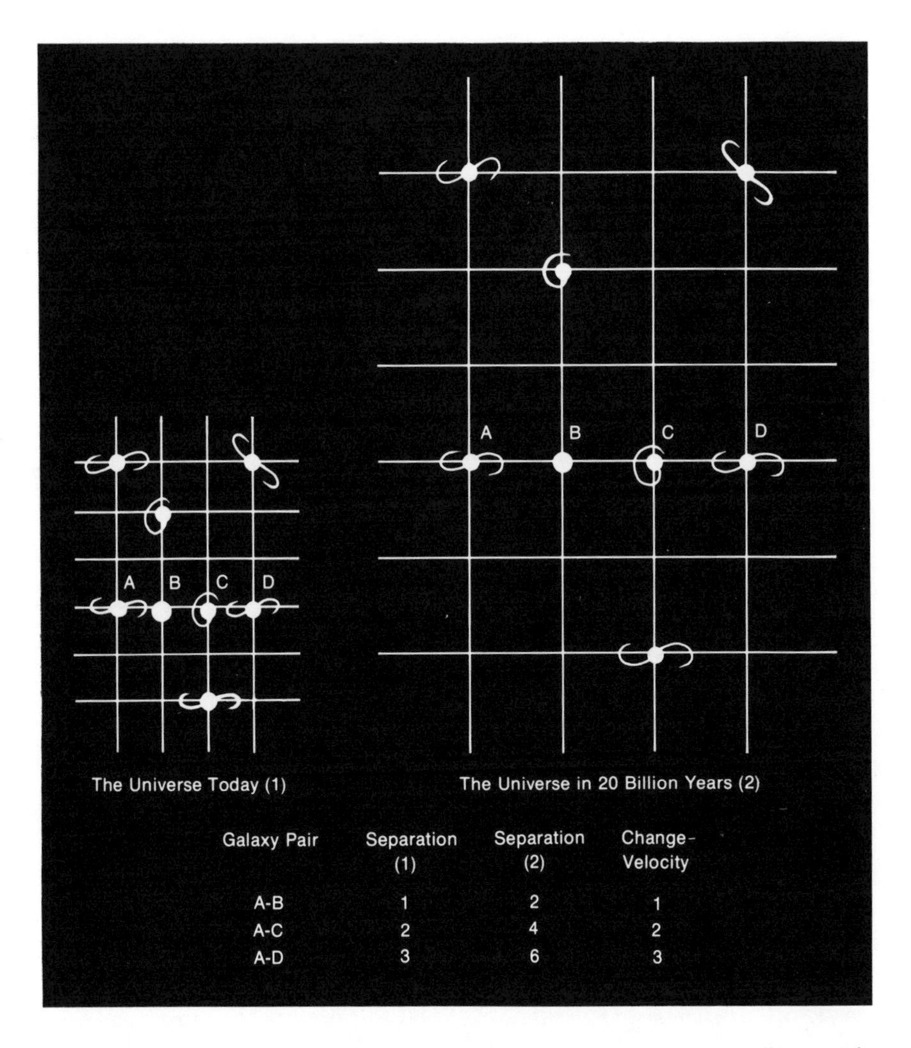

Galaxy Pair	Separation (1)	Separation (2)	Change-Velocity
A-B	1	2	1
A-C	2	4	2
A-D	3	6	3

FIGURE 13-4 As the universe expands, the separation of more distant galaxies increases faster than the separation of nearby galaxies. The expansion velocity follows Hubble's Law: Velocity is proportional to the distance.

tions made in the 1960s. He proposed that in the beginning all the galaxies in the universe were concentrated in a single lump, which he called the primeval atom. This primeval atom then exploded, flinging the galaxies off into space. While the term *primeval atom* is no longer used very often, the basic idea is still the same: in the beginning the Big Bang sent all elements in the universe moving away from one another, and this motion is still visible as the expansion of the universe.

Yet until the observations of the 1960s, this picture was not entirely satisfactory. It was just a model, and there were other ways to explain the expanding universe. People could get together and argue about which model of the universe they preferred, but all they could talk about was the esthetics of each model – which one was more philosophically pleasing. It was the introduction of observational data bearing on the early stages of the evolution of the universe that made cosmology a science. Two Big Bang remnants were discovered: the processing of one-quarter of the universe into helium, and radiation left over from this primeval explosion.

But how can we be sure that these two observations do indeed tell us about the Big Bang? Recall that the essence of the scientific process is model-building and fitting models to observations of the real world (recall Chapter 1). In cosmology, the models came first, so I turn to a detailed description of the Big Bang model, the one almost all astronomers believe in. The basic elements of this model were developed by Chushiro Hayashi and George Gamow, along with coworkers, in the 1940s and 1950s.

The beginning: the Big Bang

The story starts with a homogeneous glob of matter containing all of the substance of the universe. This glob was hot and dense, with a temperature of 10^{12} to 10^{13} K and a density of 10^{14} g/cm^3. This matter was in the form of a number of particles with exotic names: pions, hyperons, mesons, muons, neutrinos, along with the more familiar protons, neutrons, electrons, and photons. (You might wish to refer to the Preliminary section or to the Glossary if you feel a bit at sea with protons, electrons, neutrons, and photons.) When the story begins, in frame A of Figure 13-5, all these particles are in equilibrium, as they are being continually transformed through various reactions. Billions of times a second, electrons collide with their antimatter counterparts, positrons, disappearing in a flash of gamma rays. Similarly, gamma rays collide, producing electron-positron pairs. Every reaction is balanced by an inverse reaction.

What preceded the hot dense state of frame A? Some theorists have tried to answer this question, and their probings into the model world have uncovered all sorts of strange things. Apply Einstein's equations straightforwardly and you come to a singularity, just like the singularity at the center of a black hole. Both types of singularity are equally mysterious. An even stranger universe might exist: the so-called Mixmaster universe, in which the expansion rate is different in different

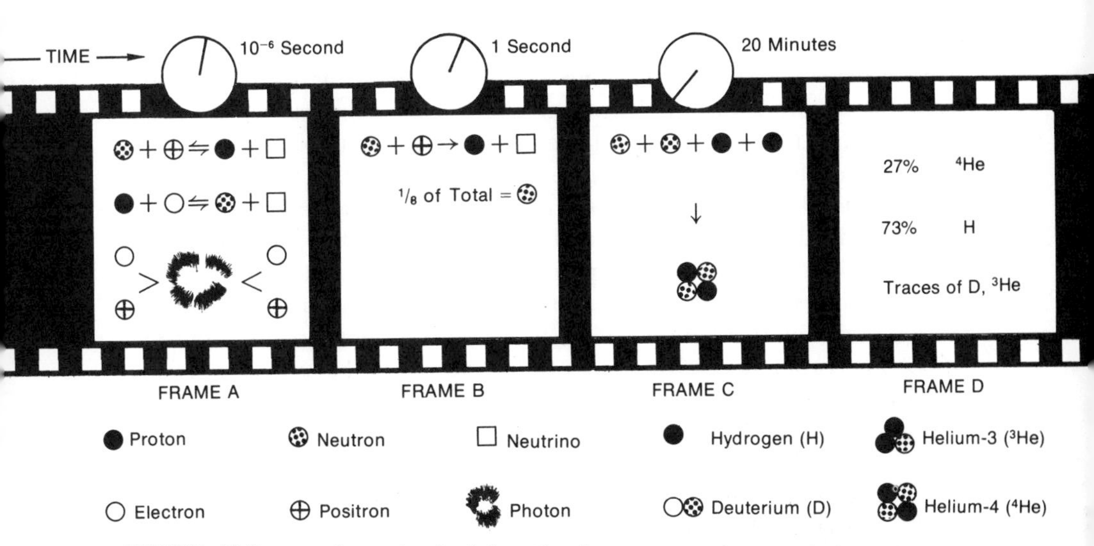

FIGURE 13-5 A movie depicting the first twenty minutes of the universe, when helium was formed. See text for details.

directions before settling down. But as with the black hole singularity, there is no way that the real world of observations can shed any light on the mysterious state preceding frame A of Figure 13-5, so I will leave it alone.

As the universe evolves, it cools and expands. Reactions no longer occur fast enough so that inverse reactions can balance them, as is shown in frame B. The positrons, being in an antimatter minority, disappear from the scene as gamma rays cannot collide and make them in appreciable quantities. The density is just too low. Once the positrons go, neutrons become stable because they no longer become protons by combining with positrons. At this one-second mark, with a density of 10^5 g/cm^3 and a temperature of 10^{10} K, the number of neutrons was frozen at one-seventh of the number of protons.

HELIUM PRODUCTION

The neutrons are important because they are responsible for the transformation of roughly one-quarter of the universe into helium. Between the one-second mark (frame B) and the 20-minute mark (frame D), the series of reactions summarized in frame C make helium. A neutron and a proton collide, stick together, and form deuterium, or heavy hydrogen. Deuterium nuclei almost always pick up another proton and neutron to become helium-4: two protons, two neutrons. A very small fraction of the deuterium does not react further and remains deuterium. The sequence stops there, because a helium nucleus cannot go anywhere by picking up another particle, proton or neutron. It would have a mass of 5, and there is no stable nucleus of mass 5. As George Gamow put it, it cannot leap over the mass 5 crevasse, so a helium nucleus finds itself at the end of the primeval element-production line.

The amount of helium produced in a hot Big Bang does not, fortunately, depend on the details of the cosmological model. Complex theoretical calculations, which carefully follow each reaction taking place in the early universe, confirm the qualitative picture sketched above: for a wide variety of initial conditions, a Big Bang universe will become roughly one-quarter helium in the first twenty minutes of its evolution. The observational confirmation of the fact that the universe is one-quarter helium thus supports the Big Bang model; I will return to the helium question in Chapter 15.

Once the helium was made, nothing very interesting happened for a long time. The universe cooled and expanded but no transformations occurred. The next interesting thing occurred much later, when the universe was some hundreds of thousands of years old. For the sake of definiteness, I refer to the age of the universe at this next milestone as 700,000 years, although the exact figure depends on the precise figures you choose for the age of the universe and the details of the cosmological model.

RADIATION

When the universe was between one second and 700,000 years old, the photons kept colliding with the matter in the universe. The principal result of these collisions was that the photons remained in balance with the matter, even though nothing changed as a result of these collisions. At the 700,000-year mark, however, the temperature of the universe had dropped to 3000 K. At this point, the universe was cool enough so that the electrons recombined with the protons in the universe, forming hydrogen atoms. (Recall from Chapter 9 the discussion of recombination.) There were no longer any electrons with which the photons could collide, so these photons simply traveled unimpeded through the universe.

The 700,000-year point is an important milestone because it provides another relation between cosmological models and the real world of observations. We can observe photons that last interacted with matter when they scattered off electrons just before recombination. Now redshifted to become radio photons, they have been detected by radio astronomers. If the radio astronomers have really detected Big Bang relics and not something else, then we can look back to the 700,000-year old universe when we look at this radiation.

The next significant event in the universe was formation of galaxies. Theory tells us little about the formation of galaxies, as no one has yet found a natural way to make galaxies in a hot Big Bang universe without invoking arbitrary assumptions. This failure to make galaxies is one potential problem for the theory, because the existence of galaxies is an undeniable observational fact. Yet it is very possible that we have just not figured out how Nature did it back then. We have some clues to when galaxy formation took place: it happened after recombination, and galaxies could not have existed in their present form until the universe was 20 million years old.

Observations of high-redshift quasars provide some information about the early universe. The quasar with the highest redshift, OQ 172 (mentioned in Chapter 7), has a redshift of 3.53 and is an occupant of the early universe. When we look at distant objects we are also looking backwards in time, into the past, as it takes light a long, long time to get from the quasar to us. Light left the quasar OQ 172 when the universe was 1.5 billion years old, or 1.5 billion years after the Big Bang. The existence of this quasar shows that something, some massive aggregation of matter, was around when the universe was a few billion years old. It is not known whether these early quasars like OQ 172 resembled galaxies.

Counts of the numbers of quasars with different redshifts (Chapter 11) indicate that the number of quasars reached a peak in the two-to-three-billion-year epoch and dropped off thereafter. You can think of this epoch as the Quasar Era, though galaxies much like the present ones could have been around. An ordinary galaxy is just not bright enough to be seen at such vast distances. The most distant galaxies that we can see have redshifts of about 0.5, indicating that the light from these galaxies left them when the universe was about seven billion years old. Infrared astronomers are currently searching for galaxies that are more distant. Other insights into the Quasar Era may come from x-ray astronomers observing background radiation, x-rays from all these distant quasars blurred together by the poor resolving power of x-ray telescopes.

From the first appearance of galaxies to the present, in the Galaxy Era, the universe evolved quietly, with no change in its overall appearance. Galaxies moved apart from each other, following the expanding universe. The galaxies themselves evolved, as their first-generation stars, which contained no metals and consisted entirely of primeval hydrogen and helium, were born and died. Subsequent generations of stars were born. In some galaxies, spirals and irregulars, the birth of generations of stars has continued, up to the present. Small stars could survive from the beginning to the present, since they, less luminous than their more massive counterparts, conserve their fuel. In our own galaxy, we see a few stars surviving from these early epochs.

Thus the Galaxy Era, though it contains many interesting events like the formation of the solar system, does not see much happening to affect the overall evolution of the universe, or cosmology. Since galaxies were formed a few billion years after the Big Bang, they have just existed. Will this always be so? Or will something more basic happen to the Universe? Let us see what theory has to tell us about the future.

The future evolution of the universe

The expansion of the universe is being continually slowed, since every galaxy in the universe is affected by the gravity of every other galaxy. Gravitational attraction is attempting to pull these galaxies together, opposing the universal expansion and causing the expansion rate to decrease. However, the expanding universe is causing the galaxies to move farther and farther apart, decreasing the gravitational attraction between them. Two alternatives seem ultimately possible. Either gravity is strong enough so that the universal expansion will slow down, stop, and become a contraction; or the expansion is sufficiently rapid that the rapidly increasing distances between the galaxies will overcome gravity, and the universe will go on expanding forever. The universe that stops expanding is called a *closed* universe, and the ever expanding one is an *open* universe. There is a borderline case, in which gravity is just strong enough that the distance between any two galaxies increases towards a limiting value, as the expansion rate is slowed toward zero by gravity, but gravity is not strong enough to bring the universe back together again.

COSMIC GEOMETRY

The present shape of the universe is also related to its future if gravity is assumed to be the only force governing its evolution. (As there is no evidence for the existence of cosmical repulsion, I shall henceforth ignore it.) The shape of the universe is somewhat difficult to comprehend; one tends to look at shapes from the outside, and we are living *within* the universe. Furthermore, you cannot visualize the shape of a three-dimensional space, the universe, unless you can think in four dimensions. I cannot think in four dimensions, much less draw in four dimensions, so you will have to be content with two-dimensional analogues of the three-dimensional universe, drawn in Figure 13-6. When you look at these things, think like a Flatlander or an ant crawling on the surface of these universes.

The closed universe is a hyperspherical universe. A hypersphere is a three-dimensional analogue of the two-dimensional surface of a sphere. We live in galaxies on the surface of this sphere and are constrained to

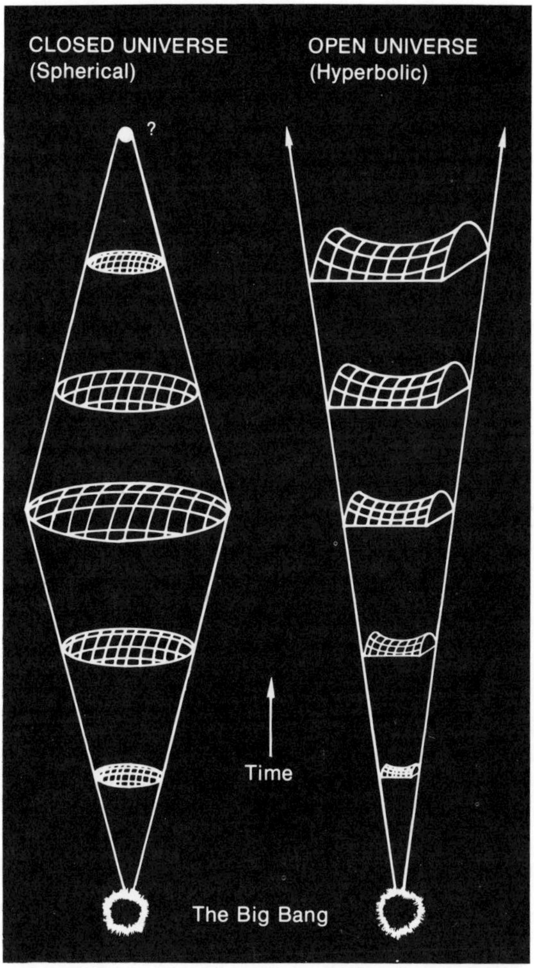

FIGURE 13-6 Open and closed universes. As the universe expands, it becomes flatter. A closed or spherical universe will stop expanding at the point of maximum size and then recontract. An open universe will go on expanding forever. The drawing shows two-dimensional analogues of three-dimensional surfaces; three-dimensional surfaces would have to be drawn on four-dimensional paper.

move along its surface. The open universe is an even stranger one; it is shaped like a saddle or hyperboloid. While the different shapes of the universe can, in principle, result in observable effects, the universe is so large that it is impossible to detect these effects with current techniques.

Neither the open universe nor the closed universe has an edge. Consider the closed, spherical universe: the surface of a sphere has no edge. The open universe is infinite and has no edge either. When people refer to the "edge" of the universe, they are referring to the edge of the *observable* universe. If you take the figure of 13 billion years for the age of the universe, anything more distant than 13 billion light-years could not be seen by us, for it would take light longer than the 13-billion-year age of the universe to get from there to here. The age of the universe limits the extent of the *observable* universe, but does not limit the extent of the universe itself, since there are galaxies beyond the observable universe that we just cannot see because they are too far away. The term *edge of the universe* is spectacular-sounding but misleading; a better term would be *limits of the observable universe.*

COSMIC EVOLUTION

What will future universes look like? As galaxies evolve, they grow old. Stars within the galaxy are evolving, becoming red giants, and dying. Some of the mass present in stars is recycled, as gas is ejected from stars in the form of planetary nebulae or supernova remnants (recall Chapters 2 and 3). Some of this ejected gas will form new stars. But the recycling is not perfect. A star forms from the interstellar gas. When it dies, it returns some of its mass to the interstellar gas, but some of its mass is interred in the stellar graveyard as a white dwarf, neutron star, or black hole. Stars in a galaxy evolve as the galaxy ages; bright blue stars disappear, returning some but not all of their mass to the interstellar medium. As the eons pass, more and more of the interstellar gas is processed through stellar evolution, finally becoming stellar corpses. There is progressively less gas to make young new stars. All that is left are the low-mass, low-luminosity stars, which burn their nuclear fuel very slowly. Eventually these stars die, too. In the end, a galaxy will consist of nothing but white dwarfs and neutron stars, which cool slowly, eventually becoming black dwarfs. In the long run, any galaxy will be nothing but lifeless stellar corpses.

This open universe does not offer any hope for rejuvenation. Galaxies will separate more and more; hence an observer in any individual galaxy will see the other galaxies become dimmer and dimmer. If his galaxy is in a cluster, he will not see the other members of the cluster recede, since it is the universe, not the gravitationally bound cluster of galaxies, that is expanding. But he will see his own galaxy and the other galaxies age, as the stars in them become older and older. In an open universe, the cosmos dies away like a clock that has been wound and left to keep ticking. Like everything in it, the universe will go through a life cycle: born in a blaze of glory in the Big Bang, live as the galaxies form and expand, each one evolving, giving birth to stars and civiliza-

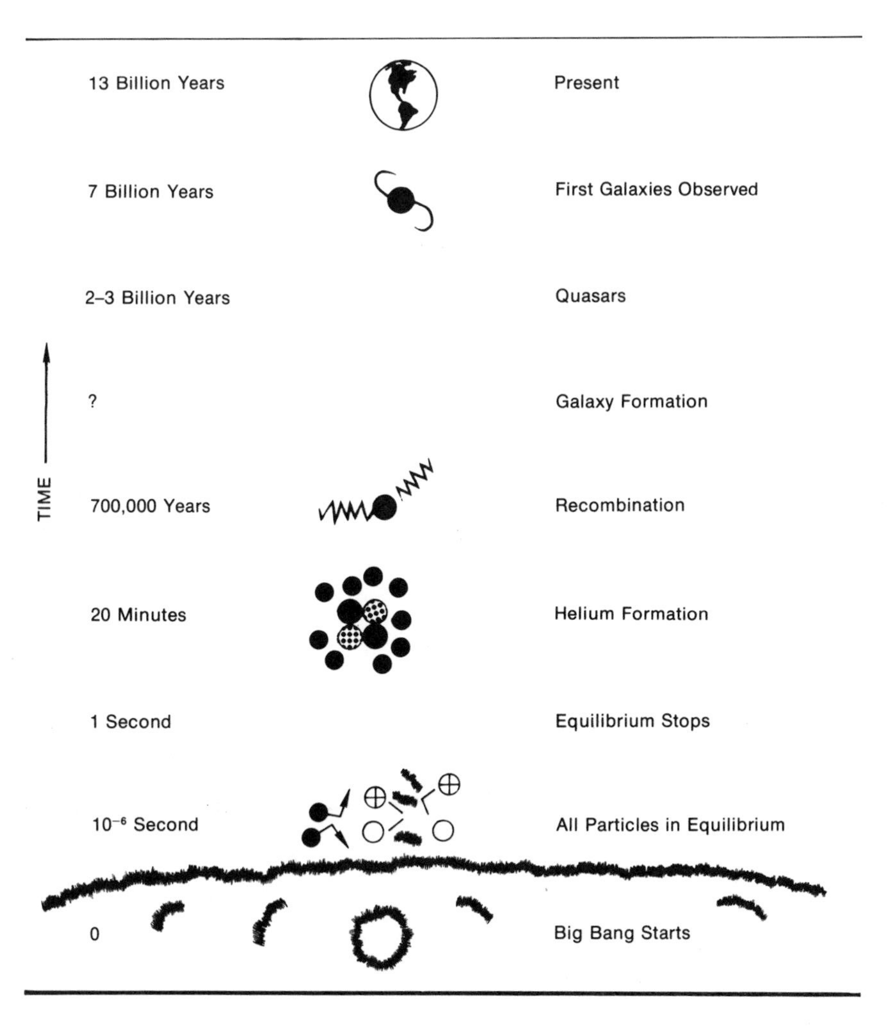

FIGURE 13-7 Review of the history of the universe.

tions, and dying as the galaxies get farther and farther apart, each one seeing its stars evolve into white dwarfs, neutron stars, and black holes.

The closed universe may offer a more esthetically promising picture. Individual galaxies will evolve just as they did in an open universe, but eventually the universe itself will offer some hope for rejuvenation. The expansion will slow down, stop, and reverse itself, turning into a contraction. The galaxies will start approaching each other again, will coalesce, and will approach a time when the universe looks as it did at the beginning, a dense glob. What happens next? Our view is limited. We cannot guess what happens at this time. Theory says that the universe will form a singular state, just like the star that made a black hole. But clearly, the theory is inadequate to describe what happens. We scientists can speculate about what will happen, but when we speculate in this way we are leaving the territory of science and are trespassing on the territory of the philosopher and the theologian.

The future of a closed universe is a mirror image of the past. As the closed universe coalesces, it becomes much hotter. The sequence described earlier will reverse itself; instead of recombination at the 700,000-year mark, there will be ionization. Helium atoms will be disrupted because the universe will become too hot to allow them to exist as nuclei. Can we obtain insight into the future by looking at the past?

No, it is not possible. Our perspective into the past meets a black curtain when we try to look beyond the stage just before helium formation (where the theoretical description presented in this chapter started). Observations can tell us that at one time the universe was hot and dense. That is all that observations available today can tell us. What happened before then is anybody's guess. Is the universe open or closed? The scenarios described above make a closed universe seem somewhat more hopeful-sounding than an open universe, but esthetic judgments have no part in such decisions. Before we take this up, we shall probe more deeply into the past, as Chapter 14 considers how old the universe really is and Chapter 15 examines the evidence for the Big Bang picture presented in this chapter. We return to the open universe vs the closed universe question in Chapter 16.

Figure 13-7 summarizes the milestones of the past history of the universe, according to the Big Bang model. These milestones are important because they provide relics of the Big Bang that can be observed today. Roughly one-quarter of the universe became helium when the universe was twenty minutes old. At the 700,000-year mark, the radiation present in the primeval fireball interacted with matter for the last time and thereafter expanded along with the rest of the universe, becoming redshifted to the radio region, where we see it today. The formation of galaxies is clouded with uncertainty; if the quasars are cosmological, they came into existence a few billion years after the Big Bang.

This chapter has been virtually all model. The next few chapters will examine the observations relevant to cosmology to see whether the model has any basis in reality.

THE COSMIC TIME SCALE

Now that we have a cosmological model, observational evidence is needed that gives us some reason to believe that the model is more than just philosophy. The theorists were first on the scene in the history of cosmology, and the cosmological models used today are essentially those that Alexander Friedmann first concocted in 1922. Most present-day work on cosmology seeks to match these models to the real world of observations.

The most obvious feature of a Big Bang cosmological model is its statement that the universe began at some definite past time. Is there any evidence for this? Objects in the universe are being formed continuously, so an examination of any one object cannot tell us how old the universe is. We can, however, seek out the oldest objects in the universe and see whether their ages point to a definite beginning of cosmic evolution. We then can compare the ages of objects in the universe with the age of the universe itself and see whether the two ages correspond.

Ages of objects in the universe

STARS

Stars are continually being born in spiral galaxies like ours. There are some clusters of stars, however, that date from the early days. These globular clusters are found in the halo of our galaxy, far from the activity in the spiral arms. We believe that the globular clusters were formed at some time in the early stages of the evolution of our galaxy; therefore if their age can be measured, we can obtain an approximate figure for the age of the galaxy.

As a star cluster ages, stars of steadily decreasing mass reach the stage where they exhaust their central hydrogen fuel and become red giants. Allan Sandage of the Hale Observatories has obtained some observations of these clusters that indicate that stars like the sun are just beginning to reach the red giant stage. Careful comparison of Sandage's observations with theoretical models for stellar evolution indicate that it takes somewhere between 10 billion and 15 billion years for such stars to exhaust their central hydrogen. The uncertainties are still substantial, for we do not know the composition of these stars accurately.

ATOMS

Some atoms are radioactive. As radioactive atoms age, they decay slowly. Nuclear reactions cause a radioactive atom to change its nature and turn into the atom of some other element. Uranium-238 will eventually become lead-206 over billions of years. We can measure the ages of atoms by checking the progress of this decay. The procedures involved here are extremely complex; a recent *Scientific American* article by David Schramm describes all the details.[1] Current research indicates that the age of the oldest atoms is between 7 billion and 15 billion years. This age refers to the age of the heavy elements produced in supernova explosions since the Big Bang; it is not possible to determine the age of the atoms produced in the Big Bang since they are not radioactive.

While stars and atoms of heavy elements are a little younger than the universe as a whole, their ages fall into the same general range of 10 billion to 15 billion years. The absence of any objects significantly older than this figure points to some finite time in the past as the beginning of our galaxy, at least. Yet our galaxy is not the universe. How does the age of the universe compare with the 10–15-billion-year age of our galaxy? If the ages are similar, we have evidence that the universe began at some definite time in the past – the time of the Big Bang.

The age of the universe

The expansion rate determines the age of the universe. Follow the expansion backward, and you come to the Big Bang – the time when all matter was aggregated. The faster the expansion, the less the time from the Big Bang, because the universe has needed less time to progress from a dense state to its present dispersed state.

Hubble's constant measures how fast the universe is expanding. This constant H is involved in Hubble's Law describing the expansion of the universe. The velocity of recession v of an extragalactic object equals Hubble's constant H times the object's distance d; thus $v = Hd$. The larger the Hubble constant, the faster the expansion rate.

The Hubble constant is a measure of the age of the universe. If it is large, the expansion proceeds rapidly and the universe is young. If it is small, the expansion is slow and the universe old. Such a statement presumes that there have been no changes in the expansion rate; the age of the universe derived in this way is called a Hubble time. Since the expansion was faster in the past, the universe is in general younger than the age the Hubble time indicates: somewhere between 60 and 100 percent of the Hubble time for reasonable values for the change in the expansion rate.

To evaluate the age of the universe numerically, we must cast the above arguments into a more precise form. For a specific example, consider the Hydra cluster of galaxies, whose recession velocity is 61,000 kilometers per second. Hubble's Law states that their distance is 61,000 divided by H, or 1220 megaparsecs, with H equal to 50 kilometers per second per megaparsec. Following this expansion backward in time, you find that it takes 6.17×10^{17} sec, or 19.6 billion years for the Hydra cluster to cover this distance if the expansion rate has been constant. A larger Hubble constant produces a faster expansion rate and a shorter Hubble time. Numerically, the Hubble constant is usually expressed in units of kilometers per second per megaparsec, and the Hubble time equals 19.6 billion years times 50 divided by H.

The Hubble constant is also used in many other areas of extragalactic astronomy. It is used, for example, to estimate the distances to quasars. Yet quasars are strange enough so that few if any aspects of the quasar story would change materially if the Hubble constant were readjusted. Its exact value is more important for cosmology.

How do you measure the Hubble constant? In principle, it is quite easy. Go and find a galaxy, measure its recessional velocity by looking for a Doppler shift in its spectrum, measure its distance, divide, and out pops the Hubble constant, as the velocity divided by the distance.

Yet there are some problems with this method. You can only measure accurate distances for nearby galaxies where stars can be seen. But the random motions of these galaxies through space swamp the expansion of the universe. These random motions are a few hundred kilometers per second. Take Messier 81, a spiral galaxy some three megaparsecs away. At this distance, the expansion of the universe amounts to 150 kilometers per second, roughly equal to the random motions. Clearly you cannot use Messier 81 to measure the Hubble constant. To measure Hubble's constant, we need to examine galaxies that are tens or, more desirable, hundreds of megaparsecs distant, since at 100 megaparsecs the universal expansion velocity of 5000 kilometers per second will overcome any local motions.

Yet at such great distances we cannot distinguish individual stars in galaxies. How, then, are we to measure distances to those galaxies? We must use galaxies in the Local Group and other nearby galaxies as steppingstones. But how do you even measure the distance to a galaxy in the Local Group? To accomplish this basic activity, we need to know how astronomical distances are measured.

Techniques of distance measurement

The elemental and most common way of determining distances in astronomy is by measuring the apparent brightness of familiar objects. As a star moves farther and farther from the earth, it becomes fainter and fainter according to an exact mathematical law. The amount of light energy that crosses the lens of your eyeball (F) falls off in the amount one divided by the square of the distance d to the star, so long as there is nothing but empty space between you and the star. (Thus F varies as $1/d^2$.) If an astronomer therefore is looking at a star that he recognizes as some familiar type of star, he knows how bright the star would be if it were some given distance from the earth. He then measures how bright it actually is and thus determines how far away it is.

This easy-sounding procedure contains one critical assumption: that the space between us and the subject star is empty. When you measure distances in the Milky Way, star clouds, which absorb light, may lie between the earth and the star under scrutiny. These star clouds absorb light and complicate the procedure. (The clouds do redden the light, as was described in Chapter 5 in connection with the distance measurement of Cygnus X-1, so you can estimate the absorption from the color change and compensate for it approximately.) As the Milky Way obscures our view of all but the closest galaxies, extragalactic work generally involves galaxies situated where our view is not obscured by the Milky Way and where interstellar absorption of starlight is not so important as it is in stellar astronomy.

An example of the application of this procedure is as follows. Suppose you want to measure the distance to the nearest star, Alpha Centauri. Begin by assuming that Alpha Centauri is identical to the sun. (Its spectrum is quite similar, so that this assumption seems reasonable.) The sun is some 5.2×10^{10} times brighter, as seen from the earth, than Alpha Centauri, because it is so much closer. Using the formula that light from an object varies as $1/d^2$ where d is the distance, you can find that the sun is 2.3×10^5 times closer $[(2.3 \times 10^5)^2 = 5.2 \times 10^{10}]$. Thus the distance to Alpha Centauri is 2.3×10^5 astronomical units, or a little more than one parsec. (Not bad for a crude approximation; the actual distance to the Alpha Centauri system is 1.3 parsecs.)

However, application of this method is awkward. Most stars are unlike the sun, so using it as a standard star, as in the formalism above, would give the wrong answer in most cases. The principle is still valid nevertheless, and partly for historical reasons, astronomers have adopted a different formalism to make such distance measurements. The brightness of a star, as seen in the sky, is measured by the apparent magnitude of the star. The brightness that that star would have if it were ten parsecs away is its absolute magnitude. The difference between the apparent and absolute magnitudes, suitably corrected for the absorption of light by interstellar clouds, is called the distance modulus. The distance modulus gives the distance of the star through application of the law relating brightness to distance. (When you cast the law into this formalism, it becomes distance modulus $= -5 + 5 \log d$, where d is in parsecs.)

Returning to the example, we find that the apparent magnitude of Alpha Centauri is 0.0 and its absolute magnitude is 4.4. In other words, it is substantially brighter than it would be if it were 10 parsecs distant (remember, large magnitudes mean faint stars; see the Preliminary section if you're confused.) The distance modulus is simply the difference between apparent and absolute magnitudes: $0.0 - (+4.4) = -4.4$, which corresponds to a distance of 1.3 parsecs, as before.

The use of absolute and apparent magnitudes casts this method of distance measurements into a much more convenient form. In summary: If the absolute magnitude of a star is known, then you know how bright it would be if it were ten parsecs away. It is easy to measure how bright it actually is. The difference between these two brightnesses, usually expressed in magnitudes, shows how much nearer or farther away the star is than the standard distance of ten parsecs.

The hard part of this method is determining the absolute magnitude of a given star. It would be nice if starlight came with a little flag on it telling what kind of star the light came from. You could then look that star up in tables and determine its absolute magnitude, for the tables would tell you the absolute magnitudes of nearby stars. (The distances to nearby stars can be determined by triangulation; see the Preliminary section.) In many cases, the spectrum of the star can act as a flag; similar stars have similar spectra. But for our purposes in this investigation, spectra are no help. Stars in the Andromeda galaxy are simply too faint to have their spectra measured. Where do we go from here?

The Local Group of galaxies

To measure the distances to relatively nearby galaxies, those within the Local Group, one needs to observe objects in these galaxies whose absolute magnitudes can be determined. Fortunately there is a type of star whose light, properly interpreted, signals its absolute magnitude as clearly as if its photons carried a little sign. These stars, called Cepheid variables, do not maintain constant brightness. Over a period that can be as short as a few days or as long as a few months, they first increase in brightness and then decline, only to increase again in a very regular periodic fashion. Henrietta Leavitt in 1912 discovered that the longer the period, the brighter the Cepheid (in general). This was a milestone of early-twentieth-century astronomy and the key step that allowed astronomers to measure the distances to galaxies. Subsequent analysis of Cepheid variables in our galaxy, whose distances can be determined by other methods, has produced a very well-defined relation between the period of a Cepheid and its absolute magnitude, provided that the location of the Cepheid in the galaxy is known. Spiral-arm Cepheids and halo Cepheids with equal periods have different absolute magnitudes, and this difference must be allowed for.

Thanks to the Cepheids, determining the distance to a relatively nearby galaxy, in which these stars can be seen, is possible. You obtain a dozen or two nights' time on a large telescope, take many photographs of the galaxy of interest, and then look for variable stars among the stars in the galaxy. You then scrutinize the photographs with a blink microscope, which allows you to look alternately at two photographs of the same object taken at different times. Any star that does not maintain constant brightness will stand out, since its image will change from one photograph to the other while the other stars in the sky remain constant. Once you find a variable, its period can be determined if you measure its brightness on all of the other photographs and see how long it takes for the star to go through a complete cycle. In principle, this process is easy, though it is time-consuming and sometimes exasperating. It is the inspiration that led to the original proposal to observe and the knowledge that the results will be worth while that makes this drudgery rewarding.

Once the hard work is done, the astronomer has only to interpret the cosmic signposts that the Cepheids are displaying. The period of pulsation of a Cepheid variable is like a label on the star, telling absolute magnitude. Apparent magnitude can be measured quite readily, and we now know the distance to the galaxy.

It is really quite remarkable that the study of a certain relatively obscure type of star — a Cepheid variable — should prove to be the key to determining the distances to the galaxies. Henrietta Leavitt had no idea that this discovery was to be the result of her research; she was just determined to know what the variables in the Magellanic Clouds were doing. This story demonstrates how unpredictable and interrelated science is. You would not normally believe that something prosaic like variable-star work would turn out to be a critical step in our understanding of distances in the universe. But science works that way. You cannot understand how the universe works without understanding its parts. Sometimes, parts of the universe may turn out to be relatively uninteresting, but some parts, like the Cepheids, can turn out to be the key pieces in solving parts of the cosmic puzzle.

Thus the distances to galaxies within the Local Group – galaxies like the Andromeda galaxy – cannot be used to measure the Hubble constant; their random motions are much larger than their movement away from us owing to the expansion of the universe. Attempts to extend the Cepheid method to more distant galaxies are futile, because Cepheids are too faint to be seen in galaxies more distant than four megaparsecs in photographs taken by ground-based telescopes. If the Large Space Telescope, a 60- to 120-incher in earth orbit, is built and launched, the Cepheid range could be extended to 40 megaparsecs – the orbiting telescope does not have to look through the obscuration and turbulence of the earth's atmosphere. But how can we extend our vision far enough to measure the Hubble constant now?

Clusters of galaxies

While Cepheids cannot be used to measure directly the distances of the more distant galaxies, the distances that Cepheids provide within the Local Group do permit us to use the nearby galaxies as steppingstones. Edwin Hubble, when he first determined the Hubble constant, noticed that the brightest stars in Local Group galaxies all seemed to have the same absolute magnitude, that is, the same luminosity. Since, as he thought, these stars could be seen in galaxies outside the nearby ones, he could use the Local Group galaxies, in which Cepheids could provide distances, as an intermediate link. He determined the distances to Local Group galaxies by using the Cepheids, determined the absolute magnitudes of the brightest stars in those galaxies from these distances, and then determined the distances to more distant galaxies by using these absolute magnitudes of brightest stars. His results, announced in 1936, indicated that the Hubble constant was equal to 526 kilometers per second per megaparsec.

This value for the Hubble constant was annoying, since the Hubble time for such a constant is 1.86 billion years, and thus Hubble's result indicated that the universe was less than 2 billion years old. There are fossils on the earth that are 3.5 billion years old, and the rocks containing these fossils were known to be this old in 1936. Subsequently, we have found moon rocks that are 4 billion years old. Clearly something must be wrong with Hubble's result — the earth cannot be older than the universe.

Hubble had made two mistakes. In 1950, Walter Baade of the Hale Observatories discovered that there were two types of Cepheids, and Hubble had been comparing halo Cepheids in the Andromeda galaxy to spiral-arm Cepheids of our own galaxy. These two types have different intrinsic brightnesses, and as a result Hubble thought that the nearby galaxies were 2.5 times closer than they really are. This error caused him to underestimate the luminosity of the brightest stars in the Local Group galaxies. In addition, what Hubble thought were bright stars in the more distant galaxies were in fact clouds of ionized hydrogen gas, which are considerably brighter than the bright stars. Allan Sandage, in rectifying these two mistakes in 1958, deduced that the Hubble constant was somewhere between 50 and 100, between one-half and one-quarter of Baade's value and five to ten times smaller than Hubble's value.

Since 1958, scientists in numerous observatories have tried to refine this value for the Hubble constant by different methods. The astronomer must be able, through a number of intermediate links, to reach beyond the nearby galaxies to more distant ones. For nearby galaxies, random motions overwhelm the expansion of the universe; even for galaxies as far away as the Virgo cluster, the nearest rich cluster of galaxies, the motions of galaxies may not be entirely uniform. You must explore the areas several tens or even hundreds of megaparsecs distant from the earth before you can be sure that you are looking at the expanding universe and not just some local motion.

The Virgo cluster of galaxies is the nearest large cluster of galaxies, so let us see if we can use it to measure H (the Hubble constant). Rene Racine, a collaborator of Sandage's at Hale Observatories, measured the magnitudes of some 2000 globular clusters near M 87, the dominant galaxy of the Virgo cluster. Assuming that the brightest of these clusters is as bright as the brightest globular clusters in the Andromeda galaxy and the Milky Way, Racine and Sandage find that the Virgo cluster is 14.8 megaparsecs distant. Combining this with the redshift of the Virgo cluster, 1136 kilometers per second, yields a Hubble constant of 77. Unfortunately there are two reasons for distrusting this estimate of H. It is not clear that you have pushed far enough away from the earth to overcome local peculiar motions, and comparing the globular clusters around a peculiar elliptical galaxy like M 87 (recall Chapter 10) with those near a spiral galaxy is treacherous.

Subsequent work, published early in 1975, indicates that this value of 14.8 megaparsecs may be too low. Sandage and G. A. Tammann compared spiral galaxies in the Virgo cluster to other closer spirals and found that the probable distance of the Virgo cluster is about 20 megaparsecs. Other investigators disagree, believing that the value is nearer 15. In any case, it would seem premature to use the Virgo cluster as a means for measuring the Hubble constant, as random velocities may interfere with measurements of the expansion of the universe, or the pure Hubble flow. (I have used the old value for the distance to the Virgo cluster earlier in this book.)

Yet the Virgo cluster can still be used as a steppingstone. Astronomers George Abell and James Eastmond at UCLA compared the Virgo cluster with two similar clusters at greater distances: the Coma cluster and the Corona Borealis cluster. Using the 14.8 megaparsec distance to the Virgo cluster, you can then determine the absolute magnitudes of the galaxies themselves. Since all three of these clusters are of the same type, it is reasonable to assume that the absolute magnitudes of galaxies in the two more distant clusters would be the same as the absolute magnitudes of comparable galaxies in the Virgo cluster. The brightest galaxy in Coma should have the same absolute magnitude as the brightest galaxy in Virgo, the next brightest galaxies should be equally comparable, and so on. Abell and Eastmond then found distances of 130 megaparsecs to the Coma cluster and 410 megaparsecs to Corona Borealis. The redshifts of these two clusters produce a Hubble constant of 53 kilometers per second per megaparsec in each case. Is this the answer? This number has been confirmed by Sandage and Tammann, in an approach combining the steps described above.

Eight steps to the Hubble constant

Sandage and Tammann use a unified approach to determining the Hubble constant, and the description of the method serves to review the basic steps on which the method depends. It goes all the way back to the beginning. The eight steps are:

1. Determine the distances to nearby stars using triangulation, and the distance to the Hyades star cluster using other geometrical methods.
2. Determine the distances to Cepheids in our own galaxy by finding clusters of stars that contain Cepheids, compare those clusters with nearby stars and the Hyades, and determine the distance to those clusters. In this way you can determine the absolute magnitude of a Cepheid of a given period.
3. Determine the distances to nearby spiral galaxies in the Local Group, the M 81 group, and the South Polar Cap group by determining the absolute magnitudes of Cepheids in these galaxies. (These last two groups are similar to the Local Group and a few megaparsecs away.)

These steps have been discussed above. Now Sandage manages to avoid using the Virgo cluster, confining his investigation to spiral galaxies:

4. From the distances to these nearby galaxies, determine the average size of the largest clouds of ionized gas in spiral galaxies. For example, the biggest such cloud in class *Sc* spiral galaxies measures 245 parsecs across.

5. As you look off into the distance, objects of the same size appear to become smaller and smaller. By comparing the angular size of gas clouds in galaxies some tens of megaparsecs distant with the sizes of gas clouds in nearby galaxies, measured in Step 4, you can then measure the distances to spiral galaxies in the 10–50-megaparsec range.

6. Now you have measured the distance to a sufficient number of spirals, so that you can determine the average absolute magnitude of a spiral galaxy of a given shape-classification, such as the *Sc*'s.

7. You now obtain spectra and redshifts of faint spiral galaxies hundreds of megaparsecs away. You can determine their distances because you know their absolute magnitudes from Step 6 and you can measure their apparent magnitudes.

8. From the redshifts and distances of many distant spiral galaxies, determine the Hubble constant H.

While this work needs finishing touches, the resulting value of H is 55 ± 7 kilometers per second per megaparsec. However, other investigators obtain different results. For example, Sidney van den Bergh compares the supernovae in distant spiral galaxies to the supernovae in our own galaxy and obtains a value of $H = 95 \pm 20$, which is in disagreement with Sandage's value, even allowing for error (the numbers after "±"). Perhaps the best that a conservative person would say is that H is probably somewhere between 50 and 100, which is the level of knowledge we had in 1958. It seems to me, though, that there is more evidence now for a value of H nearer 50 than 100. The Hubble time for $H = 100$ is 10 billion years, which is uncomfortably close to the ages of the oldest stars at 10–15 billion years (recall that the universe is younger than one Hubble time.). Consequently, I have adopted $H = 50$, a nice round number, for use in this book. Table 14-1 summarizes the different estimates of the Hubble constant. The Appendix to this chapter should enable the energetic reader to correct the distances, luminosities, and absolute magnitudes of anything beyond the Virgo cluster for possible future changes in H. I certainly hope that such changes do not occur, but in view of the difference between the present value of 50, the value of 100 generally used in the 1960s, and the value of 526 used by Hubble, it is best to view the present value of H with a little caution.

TABLE 14-1 Determinations of the Hubble Constant

INVESTIGATOR	H (km/sec/mpc)	REMARKS
Hubble, 1936	526	Original determination
Baade, 1950	200	Corrected Hubble's value for the difference between halo and spiral-arm Cepheids
Sandage, 1958	50–100	What Hubble thought were stars were really clouds of ionized gas
Racine and Sandage, 1968	77	Globular clusters in the Virgo cluster of galaxies
Abell and Eastmond, 1968	53	Coma and Corona Borealis clusters of galaxies
Van den Bergh, 1968	95	Supernovae
Sandage and Tammann, 1975	55	Eight Steps to H (see text)

This chapter has examined several different methods for determining the age of the universe. Stellar evolution, radioactive decay, and the Hubble constant all point to a universal age between 10 billion and 20 billion years. Each of these methods has some uncertainty – the logical chains leading from the observations to an age for the universe are long. The coincidence of all these results lends some support to the Big Bang idea that the universe began at a definite point in the past, some 10–20 billion years ago. In any case, the Big Bang theory has survived a test, for the existence of an object with an age considerably greater than the Hubble time would make it difficult to sustain a belief in the creation of the entire universe at one time in the past.

The Big Bang picture does more than just postulate a single date for the creation. It describes what happened in the very early universe. Now that I have shown that the idea of a universe created at a definite time in the past is compatible with observations, I shall direct your attention to other ways in which the theoretical picture can be confirmed.

APPENDIX

What if the accepted value of the Hubble constant changes again? It may be that by the time you read this book, the value of 50 kilometers per second per megaparsec that I have adopted for the value of the Hubble constant will no longer be correct. I hope this will not be the case, but I wish to hedge my bets. Changes in the Hubble constant by a factor of 2 or so will not affect the basic picture of quasars presented in Part Two, but many of the numbers in this book will need to be adjusted. The precepts below enable an energetic reader to make these adjustments, which apply only to the distances of objects more distant than the Virgo cluster of galaxies. In the formulae below, subscripts refer to the value of the Hubble constant and $h = H/50$.

Absolute magnitudes:

$$M_{\mathrm{H}} = M_{50} + 5 \log h$$
$$M_{75} = M_{50} + 0.87$$
$$M_{100} = M_{50} + 1.5$$

Luminosities:

$$L_{\mathrm{H}} = \frac{L_{50}}{h^2}$$
$$L_{75} = \frac{L_{50}}{2.25}$$
$$L_{100} = \frac{L_{50}}{4}$$

Distances:

$$D_{\mathrm{H}} = \frac{D_{50}}{h}$$

Ages: Ages of objects are proportional to the Hubble time, which scales as $1/h$.

RELICS OF THE BIG BANG

The ages of stars, atoms, and the entire universe are in the same range – 10 billion to 20 billion years. This age coincidence points to a definite time as the beginning of the evolution of the universe, but this evidence in favor of the Big Bang model is very indirect. Is there no more direct way of confirming that our universe was once in a hot, dense state?

There is. Recall the Big Bang history (Chapter 13). For all reasonable Big Bang models, roughly one-quarter of the universe is transformed from hydrogen into helium in the first twenty minutes of cosmic evolution. In addition, there was radiation in the early universe, and this radiation has not disappeared. If we can observe this primeval fireball radiation, we can confirm the Big Bang theory. The Big Bang theory makes two unequivocal statements about the present condition of the universe: it should be roughly one-quarter helium and it should be filled with the primeval fireball radiation. As this chapter will show, these theoretical ideas are confirmed.

Another cosmological model, the Steady State theory, was quite popular in the 1950s and is still found in the literature. This theory can produce an expanding universe, but if the theory is to be sustained it must find another source for the helium in the universe and for the primeval fireball radiation. Evidence exists for and against the Steady State theory, but it is difficult to reconcile the Steady State theory with observations.

Helium in the universe

In the first twenty minutes of the evolution of a Big Bang universe, almost all of the neutrons that were around at the one-second mark were incorporated in helium nuclei, producing a universe that was roughly one-quarter helium. These neutrons were around at the one-second mark because the early universe was hot enough so that protons and electrons collided, forming neutrons. It is a truly remarkable result of Big Bang models that the amount of helium produced is quite insensitive to the initial conditions, since the fractional abundance of helium is between

23 and 29 percent for all reasonable Big Bang models. We can now check the Big Bang models by asking, "Is this helium really there?"

The helium abundance problem has engaged the attention of many astronomers working in widely different fields. Some of us look for helium directly; we focus our instruments on some object that produces helium lines as helium atoms emit or absorb radiation. Others use more indirect methods; the evolution of stars containing 25 percent helium is quite different from the evolution of stars containing no helium, and comparing real stars with model stars allows one to determine the helium abundance of these stars.

Table 15-1 lists a number of different determinations of the cosmic helium abundance. This list is not meant to be exhaustive; there must be almost a hundred different helium abundance determinations in the literature. But it is representative in one important respect: of all the helium abundance determinations made, there is not one piece of evidence that indicates a helium abundance of less than 25 percent. While there are a few stars whose visible surface layers contain little or no helium, the current consensus is that the surface helium abundance in these stars is not representative of their interior abundance. Thus the helium abundance observations are consistent with the Big Bang theory, since theory says that about 25 percent of the universe became helium in the hot Big Bang.

COSMOLOGICAL HELIUM?

So far, it seems that the Big Bang theory has passed the helium test with ease. The theory says that the universe should be roughly one-quarter helium, and a wide variety of observations show that it is indeed one-quarter helium. But how meaningful is the helium test?

TABLE 15-1 Helium Abundance Determinations

TYPE OF OBJECT	HELIUM ABUNDANCE, PERCENTAGE
Gaseous Nebulae	
Milky Way	29
Small Magellanic Cloud	25
Messier 33	34
Milky Way (Radio Observations)	26
Young Stars	30–33
Globular Clusters (Indirect Methods)	30

The helium abundance test is more like a gate in an obstacle course than an accolade of approval. If we found that some part of the universe contained very little helium, the Big Bang theory would be disproved, for it states that the helium abundance of about 25 percent should be universal. Yet the existence of a high helium abundance means only that the theory is *consistent with* the observations, not *proved by* them; the Big Bang theory could be wrong and the helium could have been made somewhere else. Suppose, for example, that a supermassive star formed at the galactic center early in the galaxy's evolution and then exploded, spreading helium all over the galaxy? Then there would be a high helium abundance that did not come from the Big Bang. Thus the Big Bang theory has passed a test, but the test may not be as significant as we should like. There is a less equivocal Big Bang test.

The primeval fireball

The hot, early universe contained photons. At the 700,000-year mark, these photons interacted with matter for the last time as they bounced off the electrons that were then just about to recombine with atomic nuclei. As we see these photons now, they have been redshifted and are visible as radio waves and far-infrared (or microwave) radiation. The universe is filled with these photons. Two terms are used to describe these photons. The *microwave background* is what is observed, and *primeval fireball* refers to the radiation that should be there according to the model. The identification of the microwave background, an observed phenomenon, with the theorists' primeval fireball was a milestone of twentieth-century cosmology.

In the 1950s, when the Big Bang theory was first formulated, the originators realized that observational detection of the primeval fireball would be an excellent confirmation of the Big Bang picture. Alas, at that time no radio astronomers had equipment that was sufficiently sensitive to detect it. The primeval fireball remained in the model world until 1965, when the microwave background that almost definitely is the primeval fireball was discovered in the climax to a cosmic detective story.

DISCOVERY OF THE MICROWAVE BACKGROUND

By 1965, numerous communications satellites had been launched, and the Bell Telephone Laboratories were becoming interested in several problems involved in their operation. Bell Lab staff members Arno Penzias and Robert Wilson were given the task of finding the sources of all the static in the antenna communicating with the satellites. If they could identify these sources of static, perhaps they could eliminate them and enable Bell to use a much less powerful radio signal in operating the communications satellites.

Penzias and Wilson started by eliminating all the known sources of noise. Some radio static was emitted by the earth's upper atmosphere; and some was introduced by the transmission lines, some by the amplifier, and so forth. Yet when they finished, there was noise for which they could not account, and which they attributed to static put forth by their equipment.

Penzias and Wilson were unaware of the work of a group of Princeton scientists, Robert Dicke, P. James E. Peebles, Philip G. Roll, and David T. Wilkinson, who were building some equipment to search for the primeval fireball. Eventually the two groups made contact, and the Princeton investigators realized that the excess static in the Bell equipment did not come from the equipment but from the universe itself. In the July 1, 1965, issue of the *Astrophysical Journal,* both groups announced their results. What had begun as an experiment in telecommunications had turned into a cosmological gold mine.

CONFIRMATION

But was this microwave background really the primeval fireball? It did not come with a name tag; we had to investigate it and be sure that it was not some other form of background radiation. The key question related to the spectrum of the radiation. If this radiation is a Big Bang relic, it would have a spectrum characteristic of emission from a hot colorless object, or a blackbody spectrum. Measuring the spectrum of the radiation involves measuring its intensity at different wavelengths. Now that the background was discovered, its confirmation as the primeval fireball depended on measurements of its spectrum.

As soon as the background radiation was discovered, several radio astronomers measured its intensity at different wavelengths. Figure 15-1 shows the results; the radio measurements from the ground are shown as open circles. The observations fit the blackbody curve expected from the Big Bang theory quite well. Yet the best confirmation of the blackbody character of the radiation would be to see the turnover in the spectrum at wavelengths of 0.1 centimeter, or 1 millimeter. Such observations cannot be made from the ground, because the earth's atmosphere absorbs radiation in just the region where the spectrum starts to turn over. (Recall Figure P-4.) Instruments have to be flown above the atmosphere in a rocket or balloon.

The first rocket results, obtained by a group from Cornell in 1968, were quite a shock to theorists who were beginning to feel comfortable with the idea of a primeval fireball. The measurements, plotted as "C" in

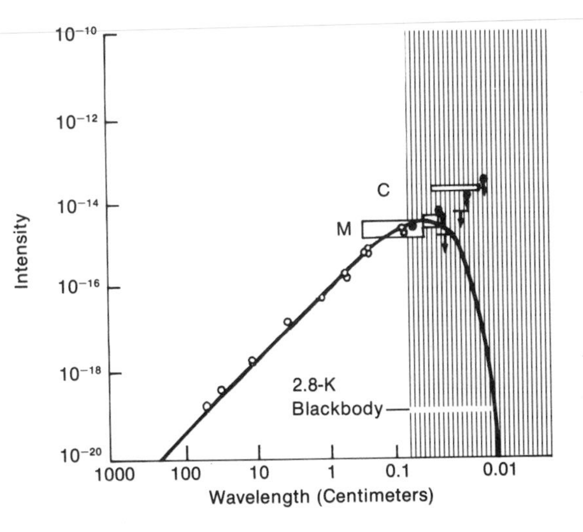

FIGURE 15-1 Observations of the background radiation. *Circles:* Ground-based observations. *Dots:* Indirect observations. M = MIT results (balloon); C = Cornell results (rocket). Vertical arrows are upper limits. (Reproduced, with permission, from "The Short-Wavelength Spectrum of the Microwave Background," by P. Thaddeus, *Annual Review of Astronomy and Astrophysics,* volume 10. Copyright © 1972 by Annual Reviews, Inc. All rights reserved.)

Figure 15-1, indicated that the instrument on board the rocket picked up thirty to one hundred times as much radiation as a black body background would have given. It was odd that some indirect measurements of the intensity of the background radiation, obtained by looking at molecules in the interstellar medium, conflicted with the Cornell results. But these measurements were indirect. An anomaly had appeared, as the background radiation no longer fitted so neatly into the blackbody, or primeval fireball, spot in the cosmic puzzle. But this anomaly did not cause people to throw out the primeval fireball hypothesis for lack of confirmation. The Cornell investigators were quite cautious about their results, more confident that their instruments were seeing something than that the radiation they were seeing was coming from the depths of the universe.

In 1969, two MIT investigators, D. Muehlner and R. Weiss, flew a telescope high up in the atmosphere on a balloon launched at Palestine, Texas. Their results from two flights are shown in Figure 15-2; the wavelength scale has been changed to make the results clearer. Their first results, in the top panel, indicated that they saw an excess amount of radiation at one millimeter, concentrated at that particular wavelength. They interpreted this radiation as emission from the upper atmosphere. Was this perhaps the radiation the Cornell investigators were observing?

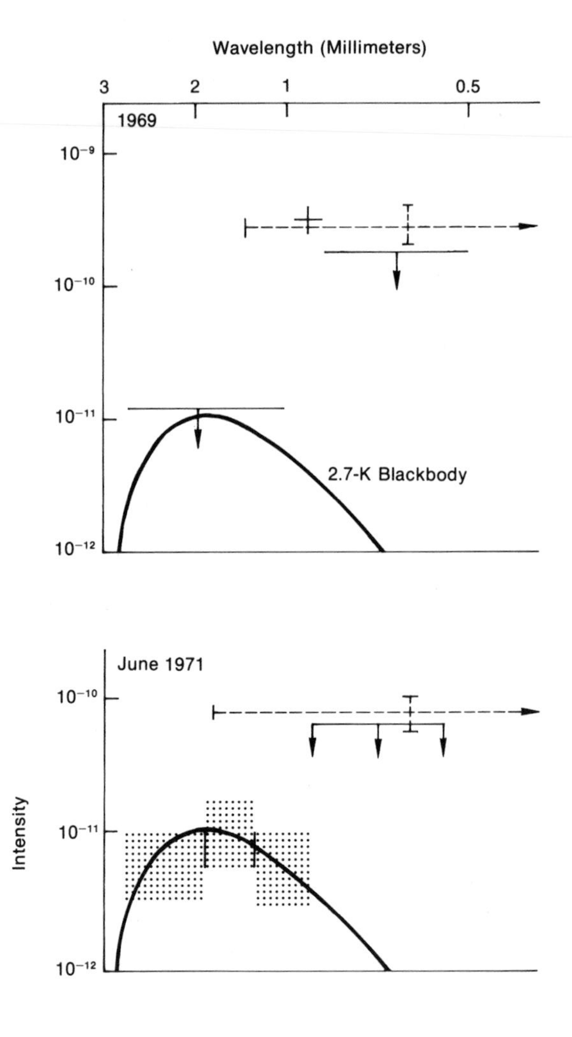

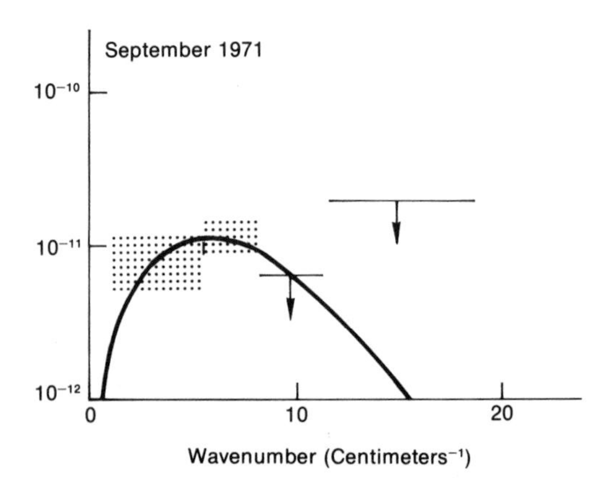

FIGURE 15-2
A detailed view of the observations of the short-wavelength background. Dotted lines: Cornell rocket results, with error bars. Solid lines and boxes: MIT results. Atmospheric corrections have been applied. (Adapted from D. Muehlner and R. Weiss, *Physical Review D* 7, p. 342.)

In June 1971, the MIT investigators flew their balloon again. They had some difficulties with their equipment, so that they were not as confident of their results as they would have liked, but they nevertheless failed to find the one-millimeter radiation that they had found two years previously. This led them to suspect that their earlier find might have been atmospheric radiation connected with some temporary atmospheric condition or that it might have been an instrumental error. In September, they repeated their experiment once again, and the results were more clearly defined, as is shown in the bottom panel of Figure 15-2. The MIT investigators were unable to detect any excess radiation. The MIT results were completely consistent with the idea that the background radiation is a blackbody. It is still not clear exactly what the Cornell investigators were seeing, if they were really seeing anything, but the current consensus is that they were not seeing background radiation in the universe.

Thus the initial reports that the background radiation might not be a blackbody, and thus not the primeval fireball, came to nothing. The anomaly went away. Such stories occur often, especially when a critical experiment is quite difficult to perform. The Cornell investigators made a very useful contribution, as they stimulated the work that led to an important confirmation of the blackbody nature of the background radiation. They were quite cautious initially, as they did not argue that they had discredited the primeval fireball interpretation of the background radiation. They simply said that they had discovered something odd, which should be checked. As the anomaly disappeared, the hypothesis that the background radiation is the primeval fireball is much stronger now than it was when they began their work.

The story is not yet finished. We should like to be able actually to detect the primeval fireball at wavelengths of less than one millimeter, not just to set upper limits. At present, there is no evidence that the background radiation is not the primeval fireball, and there is much evidence that it is; therefore identification of this background with the primeval fireball is not contested very vociferously. (Alternative explanations of this background radiation will be considered later in this chapter when I discuss the Steady State theory.)

What other ways can we use to test the hypothesis that this background radiation is the primeval fireball? If the background really comes from the Big Bang, it should be isotropic — the same in every direction, no matter where one looks. Many people have looked for possible graininess in the background radiation. If the background came from a collection of distant sources of radiation, we should expect it to be grainy and not smooth. As far as anyone can tell, the intensity of the background radiation is completely uniform, as required by the Big Bang theory.

Detection and confirmation of the primeval fireball was one of the most important discoveries of twentieth-century cosmology. As a Big Bang relic, the primeval fireball radiation has the advantage of being unequivocal. While the discovery of an apparently cosmic helium abundance is only indirect confirmation of the Big Bang picture, it is very difficult to explain the primeval fireball as anything except a relic of the Big Bang. It allows us to look back at the universe as it was at recombination, roughly 700,000 years after the Big Bang, in a way that we cannot do if we look at galaxies or even quasars.

It is even more important that the primeval fireball was predicted. Because astronomy is an observational rather than an experimental science, theoretical work generally consists of explaining what the observers have discovered. It is much more impressive and much more important for a theoretical prediction to be verified, and this happened in the case of the primeval fireball. The discovery discredited a number of other theories of cosmology, including the Steady State theory.

Demise of the Steady State theory

From the evidence presented in the last few pages, it is apparent that the Big Bang theory fits the data. But the theory is not necessarily the correct one; there may be other theories that also fit the data. Before the mid-1960s, one such theory was still alive: the Steady State theory. This theory still makes its appearance in the literature, and a few of its proponents have not given up on it. Yet the Big Bang theory has much more experimental support. If the Steady State theory had survived, it would have meant a scientific revolution, but the revolution failed, as far as we can tell now. There are still a few believers, and who knows? They may turn out to be right and the rest of us all wrong.

The Steady State theory has been good for Big Bang cosmology. Without it, the Big Bang theory would have been just accepted, not tested. Thus there are two reasons for examining the reasons that the Steady State theory is currently not believed in by most astronomers; you may run across it in some cosmology books, and it is a good foil to the Big Bang scheme. To show why the Steady State theory does not fit the data, I must turn from the real world of observations, back to the model world of cosmological theories.

THE THEORY

The basis of the Steady State theory, which Fred Hoyle, Hermann Bondi, and Thomas Gold formulated in 1948, is the old philosophical principle first stated in Ecclesiastes: "There is nothing new under the sun."

(Eccles. 1:9) This idea prevails in Western philosophy. In its mathematical form, it is termed the Perfect Cosmological Principle: the overall appearance of the universe does not change. Whenever you look at any part of the universe at any time, you will see, for example, the same number of galaxies per cubic megaparsec.

Yet the universe is expanding. How does the Steady State theory explain the expanding universe? Herein lies the theory's revolutionary character. Suppose that matter is being continuously created somewhere in space. Creation of this matter would cause a pressure that would force galaxies to move away from each other. This continuous-creation scheme violates some of the conservation laws of nineteenth-century physics, but no matter – the theory is revolutionary and thus exciting.

Figure 15-3 explains the difference between the Steady State and Big Bang views of a small part of the universe. Focus on the middle panel of the Big Bang picture first. At the present time, our small sample of space shows four galaxies, labeled *a, b, c,* and *d,* with a fifth, *e,* halfway out of the picture. In the past (lower panel), these five galaxies would be closer together. As this compaction applies to all galaxies, the lower panel contains many galaxies that cannot be seen in the middle panel for they have expanded out of our small volume of space. In the upper panel, the galaxies have moved toward the edges of our picture; the universe has expanded. Galaxy *e* has left the picture completely. The Big Bang picture is that of an evolutionary universe; the universe looks quite different in each of the three panels. The distance between galaxies changes as the universe ages.

The Steady State view of the same volume of space is shown in the right-hand side of the figure. At the present time, the Steady State universe looks exactly like the Big Bang universe. But as the Steady State universe sees time pass, the galaxies that exist now, *a, b, c,* and *d,* do move away from each other, just as they did in the Big Bang view; however, the average distance between galaxies remains the same as a new galaxy, galaxy *f,* appears. Galaxy *f* is formed from this newly created matter that is forcing the other galaxies to move away from each other. Follow the Steady State universe backward, and you see that the average distance between galaxies is still the same, since galaxies *b* and *c* did not exist ten billion years ago. The Perfect Cosmological Principle holds: On the average, the distance between galaxies does not change in time. Small differences exist, but the overall picture remains constant. "There is nothing new under the sun." It may not be the same sun, or the same galaxies, but there is nothing new about the appearance of the universe.

The Steady State theory has two philosophical consequences that, while they cannot be used to test the theory scientifically, may well have been responsible for some of its popularity. The Big Bang theory has to face a problem: Who put the Ylem, or Cosmic Egg, or whatever you

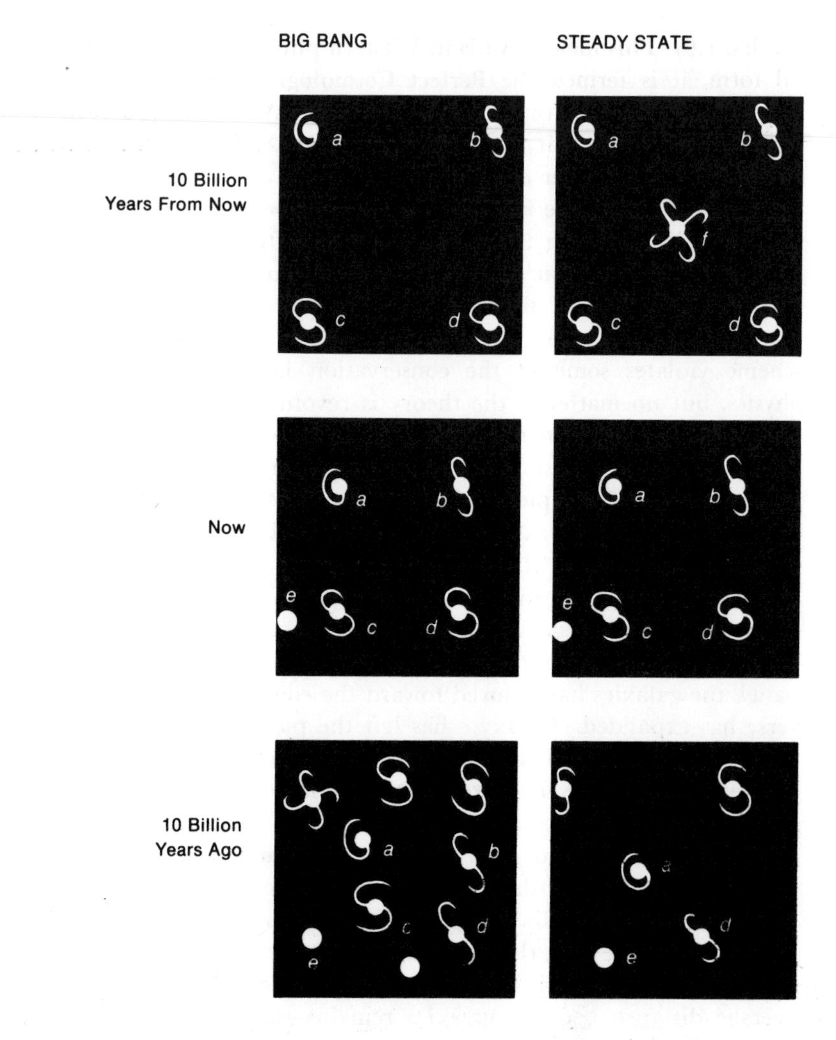

FIGURE 15-3 Two views of the evolution of the universe. See text.

wish to call it — the dense glob of matter that exploded — there in the first place? In the Big Bang theory, creation takes place at some definite time in the past. (This question can be circumvented by invoking an oscillating universe, so that the Big Bang followed the collapse of a predecessor universe, but such a model is conjectural at the present time.) The Steady State theory is also called the continuous creation theory, for matter is being created all the time. The Steady State theory thus avoids the touchy problem of origins. Furthermore, it does not produce a universe that is continually running down. These philosophical implications may explain why the Steady State theory is more popular in England than in America.

Yet philosophical meanderings are no help when it comes to testing the theory. I now return to the real world to show why the Steady State theory is no longer compatible with the observational data. It was once compatible with the real world, but one has to stretch both observations and theory now to make the theory fit the data.

COSMIC AGES

An initial motivation for the Steady State theory was a one-time discrepancy between the age of the universe and the ages of objects in it. In the late 1940s, the Big Bang theory had to face a serious problem. The Hubble time, the upper limit to the age of the universe, was then 1.8 billion years, as derived from Hubble's (1936) value of the Hubble constant. Yet the oldest rocks on the earth are known to be three billion years old, and people had begun to analyze the ages of stars and had come up with numbers like 10–20 billion years (not too different from the ages described in Chapter 14). The idea that the earth, the galaxy, and the atoms were older than the entire universe sounds absurd.

The Steady State theory avoids the age difficulty quite nicely. Any individual galaxy in the universe can be as old or as young as you like. Some galaxies are very, very old, as is shown by Figure 15-3. When it was initially proposed, it was a neat way around the embarrassing problem of ages.

The Big Bang theory has managed, with another twenty years of research on extragalactic objects, to cope with the age problem. Hubble's constant is now one-tenth as big as it was in 1950, and the universe has become ten times older. The Hubble time, the upper limit to the age of the universe, is now 19.8 billion years and there is no longer any age embarrassment.

We have not managed to knock down the Steady State theory, though; I have just shown that the ages of objects in the universe are consistent with both the Big Bang and the Steady State theories. We seek some evidence that the universe was at one time quite different from what it is now. Counts of radio sources do show the evolution of the universe, and historically they were the first pieces of evidence that cast doubts on the Steady State theory. I have not discussed this evidence until now because the interpretation of the source-count data is difficult. About all that they prove, at the present time, is that the universe does change.

COUNTING RADIO SOURCES

The strict, straightforward form of the Steady State theory is easy to refute since it makes an unequivocal prediction: the universe looks the same at all times. We can see the universe at different times, in a sense, as we look farther and farther out into space. The quasars afford us a view of the universe as it was long, long ago, since they are very distant objects and their light takes a long time to reach us. Study of the evolution of quasars shows that there were more quasars around in the early universe. This is an observational fact, as long as you agree that the redshifts of quasars are cosmological. Advocates of the Steady State theory extricate themselves from this conflict by advocating noncosmological redshifts for quasars, which have been proposed on other grounds.

There is another way to test the strict Steady State theory, independent of quasars. This test was the one that hit the Steady State theory with a blow in the 1960s. Radio astronomers counted the number of sources of cosmic radio-frequency radiation. It was assumed initially that the nearby sources were the bright ones and the distant sources the faint. There were more faint sources than there should have been for a universe that was always the same.

It soon became apparent that the interpretation of the radio-source counts was more complex than it seemed at first. You need to assume some sort of model for numerous different types of radio sources in the sky. Cosmological effects become entangled with the nature of the radio sources themselves. At one time it was hoped that radio-source counts would provide substantial information about the evolution of the universe, but these hopes have not been realized.

It is still true that the strict Steady State theory is incompatible with the existing data on the number of radio sources, as is shown in Figure 15-4. The lines are the results of a Steady State theory, indicating how many sources there should be relative to the number of sources in a flat, nonexpanding universe, with different radio brightness. It is impossible to fit any of the curves to the data. There are either too few bright sources or too few faint ones. The strict Steady State theory is simply not compatible with the data.

Present advocates of the Steady State cosmos have thus modified the theory. Both the radio-source counts and the quasar data refer only to evolution in the past ten billion years or so. Suppose that the universe goes through cycles lasting tens of billions of years, varying from epochs with many quasars and radio sources to epochs like the present one, with few quasars. These cycles do not have to be even ten billion years long, if you go along with the idea that quasars are not cosmological and hypothesize that the radio sources are nearby too. Such ideas are reasonable, even if there is no proof of them. All that the quasar and radio-

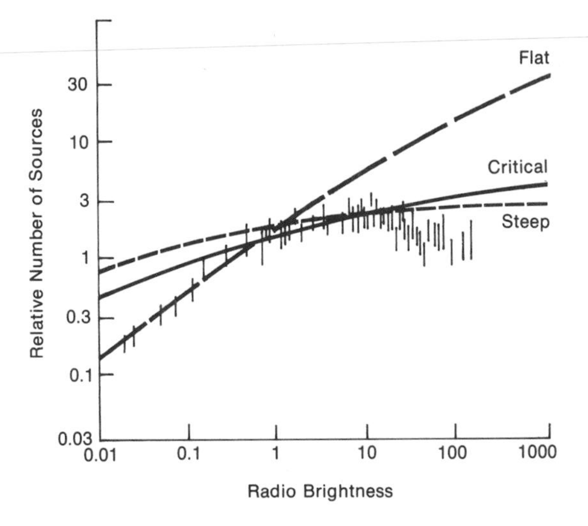

FIGURE 15-4 Source counts and the Steady State. The vertical lines are the data, the actual number of radio sources with a given brightness relative to the number in a flat, static universe. The radio brightness is in flux units 10^{-26} watt per square meter per hertz) at a frequency of 178 megahertz. The curves are theoretical results from a Steady State universe; each curve corresponds to an assumed relation between the number of radio sources and their luminosity. It is evident that no curve fits the data. (Adapted from S. von Hoerner, *Astrophysical Journal,* vol. 186, p. 750, 1973, published by the University of Chicago. Copyright by the American Astronomical Society. All rights reserved.)

source-count data prove is that the universe evolves wherever the radio sources are. The initial philosophical thrust of the Steady State theory is not so strong as it was, but the general idea persists.

The only way to rule out this revision of the Steady State theory, named the "Fluctuating Steady State," is to show that at one time the universe existed in a hot, dense state, radically different from the universe of today. The Big Bang is such a hot, dense state, and two strong pieces of evidence in this direction can be summarized.

THE BACKGROUND RADIATION

The Big Bang theory states that the universe should be filled with the primeval fireball radiation having a blackbody spectrum. As far as we can tell, this is exactly the case. The Steady State theory must find another source for this radiation. Of course, if you postulate that many galaxies are putting out radiation that just happens to mimic the blackbody background, you can save the Steady State theory. Unfortunately most radio galaxies do not radiate like blackbodies at a temperature of

3 K. Maybe you could hypothesize that there are dust grains around at just the right temperature? They would probably evaporate unless they were big enough. At this date there is no simple way to make the background radiation in the context of the Steady State model. There are complicated ways, to be sure.

THE COSMIC HELIUM ABUNDANCE

The Big Bang theory states that the universe should be between 23 and 30 percent helium. It is, so far as we can tell. The Steady State theory must provide some other source for this helium. It is somewhat easier to make helium than background radiation, but the consensus is that the helium is cosmological.

A summary of the arguments is shown in Table 15-2. You can argue that the Steady State theory is not dead yet, that it can be saved in different ways. However, it is certainly losing and losing badly. In science, particularly in astronomy, it is often possible to save a theory by patching it up to cope with the data. However, when a theory becomes more patches than original cloth it is difficult to sustain your belief in it. (Recall the mention of Occam's Razor in Chapter 1.) For this reason, few astronomers still support the Steady State theory. The Big Bang model explains the data in a much more straightforward way.

TABLE 15-2 Evidence Bearing on the Steady State Theory

	BIG BANG	STRICT STEADY STATE	HOW TO SAVE THE STEADY STATE
Source Counts			
Quasars, Cosmological Redshifts	√	X	Allow Fluctuations
Quasars, Noncosmological Redshifts	√	√	
Radio	√	X	Allow Fluctuations
Background Radiation	√	X	Find Another Source
Cosmological Helium	√	X	Make Helium Somewhere Else

In many ways, the Steady State theory was a scientific revolution that failed. Initially, in the late 1940s, there was a serious flaw, or anomaly, in the Big Bang picture; the accepted value of the Hubble constant said that the earth and some star clusters in our galaxy were older than the universe. Bad. One response to this anomaly was to look at it very hard and see whether it would go away. It did, with the Hubble constant revision. Another response, however, was to create another theory. This other theory was revolutionary, as it would have taken us away from our picture of an evolving universe, back to a static universe. However, the theory has had difficulty in explaining the observations. Counts of quasars and radio sources indicate that the universe has evolved somewhat, and the Fluctuating Steady State theory had to be introduced to cope with these observations. However, the Steady State idea still has to find an alternative explanation for two observations that indicate that at one time the universe existed in a hot, dense state: the primeval fireball radiation and the cosmic helium abundance. Such explanations are not easy to come by, and the Steady State theory is not accepted by many people these days.

Yet even a failed scientific revolution, the Steady State theory, was good for cosmology. Without it, the expanding universe of the 1930s would have remained alone on center stage, and it would have been accepted rather than tested. The radio-source counts, which had to be made with a large, complex radio telescope, would never have been made, and the telescope (which discovered a number of other exciting phenomena like radio galaxies) might not even have been constructed. People would have been much more willing to accept the background radiation as a Big Bang relic without testing it. Observational cosmology would be a much poorer science without the Steady State theory, in the same way that the discordant-redshift idea has illuminated the study of quasars.

THE FUTURE OF THE UNIVERSE

It is much easier to investigate the past than the future. The theory that the universe began as a dense glob that exploded, making helium and background radiation in the process, is strongly supported by the evidence. The present expansion is a result of this past explosion. But what happens now? Will the expansion continue forever, or will it slow down, stop, and reverse itself? It seems foolish to try to predict the future, but scientists have an advantage over historians, for the scientists know the forces that govern the future evolution of the universe. Gravity is the force that will cause the expansion to slow down. Is gravity strong enough to stop the expansion?

There are some philosophical reasons for hoping that the universe will be closed or open, depending on your point of view. Some people find the idea of an oscillating, recycling universe more appealing than the idea of a universe expanding forever, with galaxies getting farther and farther apart. Some find the oscillating universe idea less acceptable than the open one. But see what rational investigation has to say.

You can approach the decision between the alternative futures directly or indirectly. The direct way is to count up all the mass in the universe and see whether its gravitational attraction is sufficient to halt the expansion. Alternatively, you can look backward in time by looking at far-distant galaxies whose light has taken a long time to reach us and see whether you can detect any changes in the expansion rate. A third approach, also indirect, is to look for additional effects caused by the amount of matter in the universe.

The verdict is still open. The literature on research in astronomy reflects changing trends. A few years ago, the preponderance of opinion favored a closed universe, where expansion slows down, stops, and turns into a contraction. In the last two years, the trend has reversed, and is currently toward an open universe, one that expands forever. What these changing trends really mean is that the evidence is not at all complete. This chapter provides a useful framework for understanding how astronomers are attempting to answer the question, How will the universe end?

I consider the direct method of attacking this question first: counting up all the mass in the universe and ascertaining whether there is enough mass there to close the universe. First, we must know how much mass is needed.

Escape velocity

Gravity is the force that will cause a closed universe to stop expanding and start contracting. Any pair of galaxies that are flying apart from each other feel a gravitational drag on expansion, because gravity is trying to pull the galaxies together. As the galaxies separate more and more completely, however, this gravitational drag weakens as gravity decreases with decreasing distance. Whether gravity or expansion will win depends on the speed of the expanding galaxies. If the galaxies are traveling fast enough, they will overcome gravity and continue to separate from each other, forever. If the gravitational force is too strong and the galaxies are traveling too slowly, their speed will be insufficient and the moving galaxies will slow down, stop, and come together again.

A more familiar example of two objects that show the conflict between gravity and motion comes from the space program. If a rocket is to reach the moon, it must escape from the earth's gravitational field. It can do so only if it is traveling fast enough. An insufficiently fast rocket will eventually slow, stop, and return to earth, as the first few unmanned American moon shots did (see Figure 16-1). A faster rocket can overcome gravity and reach the moon. The speed at which a rocket must be traveling to escape the earth's gravity is called escape velocity, and for the earth it is equal to 11.19 kilometers per second (about 7 miles per second, or 25,030 miles per hour). Thus the cosmological problem can be restated: Are the galaxies traveling at escape velocity?

The magnitude of the escape velocity for any object or any collection of objects is determined by the strength of the gravitational forces that the objects produce. The gravitational force between any two objects is proportional to the product of their masses and inversely proportional to the square of the distance between them. Thus the more massive these two objects, the stronger the gravitational force and the larger the escape velocity. The farther apart the two objects, the weaker the gravitational force and the smaller the escape velocity. Hence, to determine whether the universe is open or closed, we must measure its mass, its speed, and its size.

Unfortunately, it is not easy to measure directly the mass, speed, and size of the universe. However, the problem can be cast into an alternative framework.[1] The Hubble constant is a rough measure of how

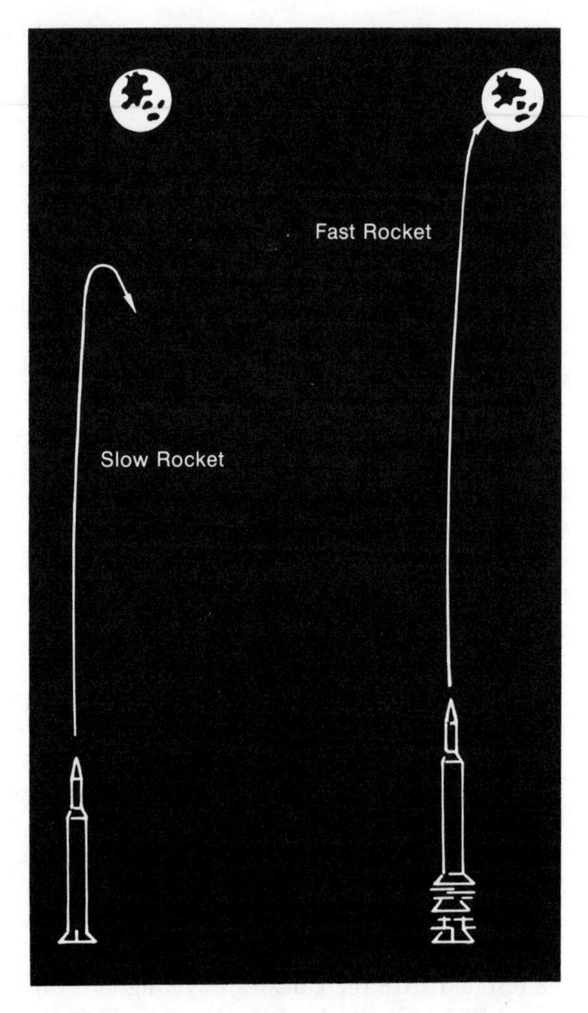

FIGURE 16-1 If you are trying to send a rocket to the moon, you must escape from the earth's gravitational field. If your rocket is traveling too slowly, at less than escape velocity, it won't reach the moon (left). It must be traveling faster than escape velocity (right).

fast the universe is expanding relative to its size. The values for the mass and size of the universe can be combined to give its density, or mass per unit volume. You can measure the density simply by counting, by looking at a few cubic megaparsecs of space and determining what is there. The cosmological question, What will happen to the Universe? can be phrased in an alternative way that is easier to answer quantitatively – Is the mean density of matter in the universe sufficient to provide a large enough gravitational force to cause the universe to stop expanding? This critical density is generally referred to as the closure density, since it is the density required to make a closed universe, in which expansion will stop and turn into a contraction. Numerically the closure density is $4.7 \times 10^{-30}\ h^2\ \mathrm{g/cm^3}$, where h is the Hubble constant divided by 50 kilometers per second per megaparsec. Since the current best value of the Hubble constant is 50, h equals 1 and the closure density is 4.7×10^{-30}, or one particle per cubic meter – one atom in a volume of space the size of a typical desk. It doesn't take much mass to close the universe, but the universe is quite empty.

The density method of deciding whether the universe is closed or open can really only show that the universe is closed. If we find enough mass to close the universe, fine; the universe is closed. If we fail to find this mass, we cannot be sure that the universe is open, for we may have overlooked some mass or it may be invisible. Let's start counting to see whether we can reach the magic figure – the closure density of 4.7×10^{-30} $\mathrm{g/cm^3}$.

The mean mass density of the universe

The universe contains many types of objects: galaxies, clusters of galaxies, and probably intergalactic stars. There may also be intergalactic gas, and possibly other, stranger constituents like black holes and planet-sized objects in intergalactic space. In this section each class of object will be considered individually, and the fraction of the closure density present in each form will be determined. (If you're anxious to find out what the answer is, skip ahead and look at Figure 16-5, which summarizes the results.) In addition to providing information about the future of the universe, this section should be a useful summary about just what is out there.

VISIBLE GALAXIES

Galaxies are the most obvious components of the universe. It is easy, in principle, to determine how much mass is contained in galaxies; you just look at wide-field photographs of the sky, such as those taken with the Palomar 48-inch Schmidt telescope, and count galaxies. Tedious, but not difficult. You determine the brightness of each galaxy and roughly estimate its distance, so you can determine the number of galaxies per unit volume. It turns out that there are between one and three large galaxies per 100 cubic megaparsecs, so the distance to the nearest large galaxy is, on the average, between three and five megaparsecs.

We then need to convert our number counts into a mass density, or mass per unit volume, by using standard values for the masses of galaxies. It is this step that is least certain; allowing for a reasonable margin of error, between 1 percent and 4 percent of the closure density exists in the form of visible galaxies. If massive halos surround galaxies (recall Chapter 7), we may be underestimating the mass in galaxies. We may also be missing very small galaxies that are not bright enough to stand out against the background of the night sky.

There are two ways to find out whether we are missing a cosmologically interesting amount of mass by neglecting these (in some cases hypothetical) forms of matter. We can search for the cosmic light—background light in the universe—contributed by this stuff, or we can see whether they exist in clusters of galaxies. The second method is more limited since it cannot detect this mass outside of clusters of galaxies.

THE COSMIC LIGHT

Many people have spent a lot of time and effort measuring the light of the night sky. It is a frustrating business, for once you eliminate the light of terrestrial activities you have a long way to go before finding anything of cosmological interest. Much of the background light in the night sky comes from light sources that illuminate dust grains, a ubiquitous part of the universal landscape. What happens is that some source of light shines on a dust grain that then reflects this light back to the telescope. A large contribution to the light of the night sky comes from illumination by city lights of dust in the earth's atmosphere, and this source of sky brightness is becoming increasingly annoying to astronomers. More interesting sources of night-sky emission are zodiacal light, which comes from sunlight illuminating interplanetary dust grains, and diffuse galactic light, starlight reflected by interstellar dust grains. Take out the light from the Milky Way too, and what remains is the cosmic light—light from galaxies, clusters of galaxies, and the baby galaxies that might have been overlooked in counting the mass density in the form of galaxies. The massive galactic halos would also contribute to this light.

The cosmic light has not been observed, but interesting upper limits have been set on its intensity. The observations indicate that it must be less than three to ten times the light from visible galaxies. You need to guess how massive the stars that contribute to this light are, so you can translate this limit on the luminosity into a limit on the mass. It turns out that the *maximum* mass that the matter giving the possible cosmic light could contribute is 2 percent of the closure density times the mass-to-light ratio of the stars that contribute to it. The mass-to-light ratio of a star is its mass divided by its luminosity in solar units. (Massive stars, which are very bright, have mass-to-light ratios *smaller* than 1; stars less massive than the sun have ratios *greater* than 1, for stellar luminosities are more variable than stellar masses.) Thus the closure density can be reached only by assuming that the cosmic light is there and is as bright as the upper limits that observers have set on it. It must also be assumed that the cosmic light is contributed by very small stars, with an average mass of 0.1 solar mass and a luminosity of 2×10^{-3} times the luminosity of the sun. We see no great number of stars in our immediate vicinity, but small stars are hard to see.

To summarize, if the cosmic light exists, it might possibly provide the mass required to close the universe. An improvement in the upper limits to the intensity of the cosmic light or an improved understanding of the probable masses of stars found outside of galaxies is needed before any positive statements can be made.

CLUSTERS OF GALAXIES

The disadvantage of trying to determine cosmologically interesting information from the cosmic light is that you observe light, not mass. A more direct mass measurement comes from studies of clusters of galaxies. If a cluster of galaxies is to be stable, the energy of motion of its component galaxies must be balanced by their gravitational energy. If the cluster is to stay together, the galaxies must be traveling at less than escape velocity from the cluster. It is generally believed that clusters of galaxies were formed only a few billion years after the Big Bang, when the galaxies were formed, so that their continued existence from then until now is good evidence that these clusters are bound together. As the escape velocity is determined by the cluster's total mass among other things, you can measure the mass of the cluster by observing the velocities of the galaxies in the cluster. It turns out that if the cluster is stable, the galaxies should be traveling at 71 percent of the escape velocity on the average; so you can directly measure the escape velocity of the cluster. Knowing the size of the cluster allows you to measure the cluster's total mass. Comparing the mass of the cluster with the masses of the individual galaxies tells you whether you are missing something or not.

Several clusters have been studied in this way, and the most thoroughly investigated is the Coma cluster, named for the constellation Coma Berenices, which contains this cluster of galaxies (Figure 16-2). If you attribute normal masses to the galaxies in the Coma cluster, the individual galaxies are traveling at two or three times escape velocity (Figure 16-3). If the cluster really were flying apart, it would no longer be visible as a cluster. We must be missing some mass.

This missing mass in the Coma cluster has been searched for and

FIGURE 16-2
The Coma cluster of galaxies. (Hale Observatories photograph.)

not found, by looking for its light. It may well be in the form of massive halos of low-mass, low-luminosity stars surrounding the individual galaxies, or it may be in some other form. But suppose it's really there; it may be that our understanding of the dynamics of the motions of galaxies in clusters of galaxies are not what we think. Even if you accept the Coma results at face value, they indicate that galaxies are six times as massive as previously thought, or that some form of mass exists within clusters of galaxies that is five times the mass of the galaxies in the cluster. Even if you extend these results to all galaxies, the mass density only reaches 12 percent of the closure density.

To summarize, the amount of mass contained in the form of galaxies is very poorly known at present. A straightforward approach indicates that the mass in visible galaxies is between 1 percent and 4 percent of the mass needed to close the universe. If you assume that the extra mass needed to bind the Coma cluster is present in all clusters of galaxies, from rich clusters to small groups, you still get only 12 percent of the closure density. Observations of the upper limit to the cosmic light indicate that intergalactic stars, baby galaxies, and massive galactic halos can close the universe only if they are composed of very small, low-mass, low-luminosity stars. The evidence indicates that the mass necessary to close the universe has not been found in the form of galaxies, and there are indications that the necessary mass is not there in this form — it is somewhat implausible that there are very large numbers of low-mass, low-luminosity stars in the universe.

So far, we have been considering stars, which can be seen because they shine. Suppose the mass that will close the universe is in the form of intergalactic gas. Is there any way that we can find it?

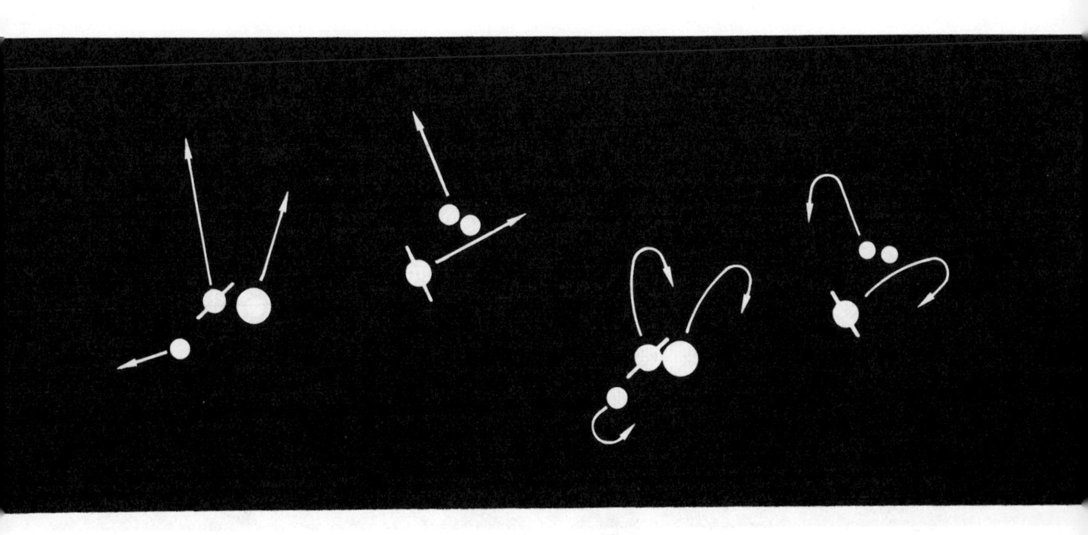

FIGURE 16-3 If the galaxies in the Coma cluster have mass normally attributed to galaxies of their type, the cluster will fly apart *(left panel)*; but if there is extra mass in the cluster, it will stay together *(right panel)*.

The intergalactic medium

Within our galaxy, the space between the stars is filled with a tenuous gas called the interstellar medium. Sometimes this gas tends to clump in clouds, and you can see these clouds as the dark dust lanes in the Milky Way photograph, Figure 7-1. It is conceivable that the space between galaxies is filled with gas; such gas has been called the intergalactic medium, even though its existence is still hypothetical. Perhaps it is here that people who want to see a closed universe will find the necessary mass. In fact, it would be quite surprising if the process of galaxy formation were so efficient that all of the matter in the universe found its way into galaxies with none left over. An astronomer searching for this gas must imagine what form it might be in and then seek it observationally. If the intergalactic medium is there, it is likely to be mostly hydrogen.

ABSENCE OF NEUTRAL HYDROGEN

The intergalactic medium is not in the form of neutral hydrogen. Neutral hydrogen would absorb radiation from quasars in a part of the spectrum that the quasars' high redshift allows us to photograph from the ground (Figure 16–4). James Gunn and Bruce Peterson, then as graduate students at CalTech, were the first to apply this test to a quasar

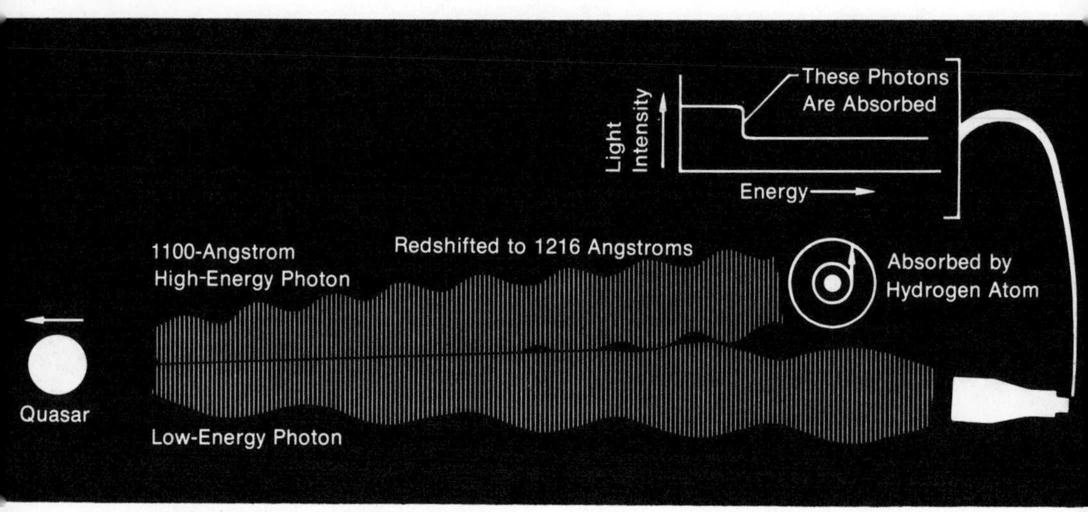

FIGURE 16-4 High-energy photons emitted by a quasar would be redshifted to 1216 angstroms and absorbed by hydrogen atoms in the intergalactic gas if the gas were there in the form of neutral hydrogen.

spectrum, and they found that there was no detectable atomic hydrogen in the intergalactic medium. This test is remarkably sensitive; if there were only 3×10^{-7} of the closure density in the form of atomic hydrogen, they would have seen it. Our search for enough matter to close the universe does not end here.

INTERGALACTIC IONIZED HYDROGEN

If the intergalactic medium is not neutral, it must be ionized if it is there at all. How could we find such a medium? In a hot ionized gas, electrons and protons fly around at high speeds. As an electron zips by a proton, its path is altered slightly as the proton's electrical force acts on it. The electron is slowed by the proton, losing some energy of motion. This energy is emitted in the form of a high-energy photon, and it is these photons that we hope to find. X-ray astronomers may have detected this radiation in the form of an x-ray background radiation that fills the universe. The background is definitely there, but its source is unknown.

Although the x-ray background could come from the intergalactic medium, recent work has shown that there is not too much hope for a closed universe here. If the intergalactic medium is massive enough to close the universe, this medium must be spread uniformly through intergalactic space or else it will produce too strong an x-ray background. Special models are needed to make a massive intergalactic medium consistent with the x-ray observations. And the x-ray background could have nothing whatever to do with the intergalactic medium. Closed-universe types, take heart; here may be the mass that you seek.

NONLUMINOUS MATTER

People who like to look really hard for the mass needed to close the universe point out a rather sobering thought for astronomers: there are many things we cannot detect in intergalactic space. We cannot see any large object outside the solar system unless it emits light. Maybe space is full of dark objects that could contribute enough mass to close the universe. What could they be? Planets, dead white dwarfs, black holes – we could never find them if they were there.

Yet when you think about it for a while, this problem looks somewhat less serious. These nonluminous objects could not be small grains, since they would act like smoke in a smoke-filled room and absorb light from distant galaxies. Anything smaller than a good-sized city would evaporate in the 13 billion years that have passed since the Big Bang. (Smaller objects in clusters of galaxies could survive, but we have seen that the mass in clusters of galaxies is not enough to close the universe.) It is hard to see how dead intergalactic stars (black holes and so on) could be produced in sufficient numbers – remember, we need about fifty times the mass in visible galaxies to close the universe. In summary, you cannot rule out the presence of a cosmologically interesting amount of nonluminous matter, but it is difficult to understand where this stuff might come from.

THE MASS DENSITY OF THE UNIVERSE

Figure 16-5 summarizes the immediately preceding thoughts. This long excursion through the matter content of the universe has provided much information about what is present in the universe and what is not, although it has not brought us much closer to a direct solution of the cosmological problem. Visible galaxies do not provide enough mass to close the universe. Other luminous material in the form of massive galactic halos or baby galaxies or intergalactic stars could provide the necessary mass only if it were concentrated in the form of low-mass stars; this condition, while possible, is not very plausible. The extra mass present in clusters of galaxies, which may be there to prevent the clusters of galaxies from dissipating, is insufficient to close the universe. The most dramatic limit on a possible constituent of the universe is the limit on the amount of mass present in an un-ionized intergalactic medium, which provides less than 7×10^{-5} (molecules) or 3×10^{-7} (atoms) of the closure density. The best hope for the mass necessary to close the uni-

FORM	DENSITY (g/cm^3)	FRACTION OF CLOSURE DENSITY
 Galaxies	9×10^{-32}	0.02
 Cosmic Light from Galaxy Halos, Intergalactic Stars, etc.	Less than 9×10^{-32} (M/L)	Less than 0.02 (*M/L*); *M/L* Must Be 50 If This Is to Close the Universe
 Extra Mass in Clusters	7.2×10^{-31}	0.15 If the Coma Cluster Is Typical
 Neutral Hydrogen	Less than 1.5×10^{-36}	Less than 3×10^{-7}
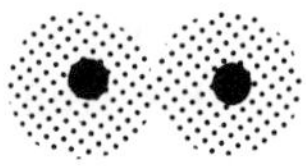 Molecular Hydrogen	Less than 3×10^{-34}	Less than 7×10^{-5}
 Intergalactic Ionized Gas	?	Maybe 1? Maybe 0?
Closure Density	4.7×10^{-30}	

FIGURE 16-5 Matter in the universe.

HOW YOU FIND IT	SOURCES OF ERROR	DEPENDENCE ON H
Count Galaxies	Don't Know Galaxy Masses Very Well	—
Look for the Cosmic Light	What Stars, with What Masses, Make Up the Cosmic Light (If It's There)	—
Analyze Motions of Galaxies in Clusters	Limitations of Our Understanding of Cluster Dynamics	$1/H$
Look at Quasar Spectra	Are Quasars Cosmological?	$1/H$
X-Ray Background	What Is the X-Ray Background? Maybe It's Something Else	

verse lies with an ionized intergalactic plasma, and future work with x-ray telescopes will be very helpful in unraveling the nature of the x-ray background.*

This search for the mass necessary to close the universe, which resembles in some ways the search for the Holy Grail in the King Arthur romances, was undertaken because of some other results from indirect lines of investigation that pronounced a closed universe. In the early 1960s, these results caused people to wish to look for the mass needed to close the universe. In the last few years, however, some doubt was cast on these results and some others came to light. So, frustrated in an attempt to close the universe by looking for enough mass, let us consider some indirect methods.

Changes in the expansion rate

If the universe is ever to stop expanding and recontract, the expansion must be always slowing down. Is there any way that this slowdown can be detected? When we look out into space, we are also looking backward into time. Thus, if we can measure the redshifts of galaxies sufficiently far into the past, we can perhaps see whether the expansion was faster in the past. Figure 16-6 shows how. Suppose that the universe were practically empty, with a mass density of much less than the closure density. With such little mass, the expansion rate would be constant at all times. When you looked outward into space, the redshifts would fall along the line labeled "no slowdown." If the expansion is slowing down significantly, as it heads towards the moment when it turns into a contraction, it would have been more rapid in the past, producing greater redshifts. The relation between distance and redshift would follow the line labeled "slowdown," for when we look at more distant galaxies we are looking at the universe as it was in very early times, long ago. The distant galaxies, or the galaxies that we see as they were in the distant past, would be more highly redshifted as they are in the graph. This is the reason that the curve bends upward if the universal expansion rate is changing.

If we can turn to the real world and put some observations on the distance-redshift relation of the left panel of Figure 16-6, we can see whether the expansion of the universe is really slowing down. Yet the

* An independently written review of the mass density question, which is more positive about an open universe than mine, is "An Unbound Universe?" by J. R. Gott III, J. E. Gunn, D. N. Schramm, and B. M. Tinsley, *Astrophysical Journal* 194 (1974), pp. 543–554.

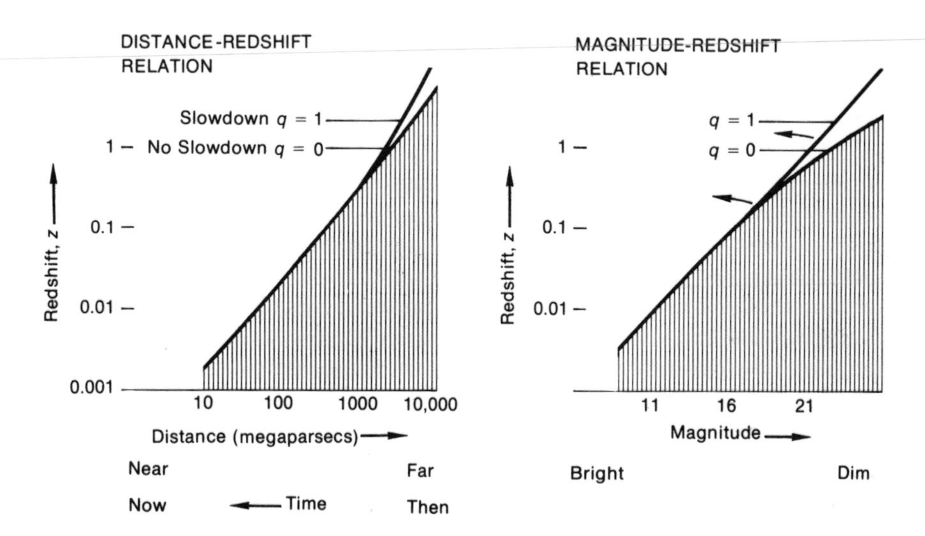

FIGURE 16-6 Is the expansion of the universe slowing down? *Left Panel:* It the expansion is slowing down ($q=1$), the redshifts of galaxies will be larger in the past or at great distances than they would be if the expansion rate did not change ($q=0$). *Right Panel:* For an object whose luminosity is constant, observed magnitude depends on its distance from observer. Near galaxies will be bright and distant galaxies will be dim. The distance-redshift relation of the left panel can thus be transformed into the magnitude-redshift relation of the right panel. The arrows show what happens if galaxies were more luminous in the past.

distances in the figure are so vast that they cannot be measured directly. The only way to allow observations to enter the picture is to suppose that there are galaxies whose luminosity is the same everywhere in the universe. The apparent brightness, or magnitude, of such galaxies (called standard candles) would thus be a good indicator of their distance. An analogous situation might exist on a highway. If all automobile headlights were equally luminous, you could determine the distances of cars quite accurately by looking at the brightness of headlights; the dim ones would be far away and the bright ones close. Thus with the assumption that standard candles exist in the universe, we can put observations on the magnitude-redshift relation of the right panel of Figure 16-6 to see whether the universal expansion slows down, to see whether the universe is closed.

The different curves in Figure 16-6 are labeled with the number q, which is called the deceleration parameter. What this number measures is how rapidly the universe is slowing down; very roughly it measures the change of the Hubble constant, fractionally, in one Hubble time. The number q is equal to one-half the ratio of the actual density of the universe to the closure density, so that if q is bigger than 0.5 the universe is closed, and if q is smaller than 0.5 the universe is open. If the actual density of the universe is the density in visible galaxies, 0.02 times the closure density, then the value of q is 0.01 and the universe is open. If standard candles exist, all you have to do is make observations of their magnitudes and redshifts, enter the figures into the right-hand panel of Figure 16-6, and see what sort of a universe we live in.

In the 1960s, observational cosmologists discovered that the brightest galaxies in rich clusters of galaxies tended to have the same luminosity wherever they were found in space. One way to visualize this is to look at the lower left-hand corner of the graphs in Figure 16-6. No matter what universe we live in, a standard candle with a redshift z of less than 0.1 should fall along the straight line, since the change in the expansion rate of the universe has little effect here. The fact that the observed magnitudes of the brightest elliptical galaxies in rich clusters fall along this line, as is shown in Figure 16-7, supports this view, that the brightest galaxies in rich clusters have the same luminosity everywhere and are thus standard candles.

There are some problems, though, which are indicated by the complicated label on the horizontal axis of Figure 16-7. Redshifts of galaxies are straightforward to measure. You just obtain a spectrum, recognize the lines in it, measure their wavelength, see how much they

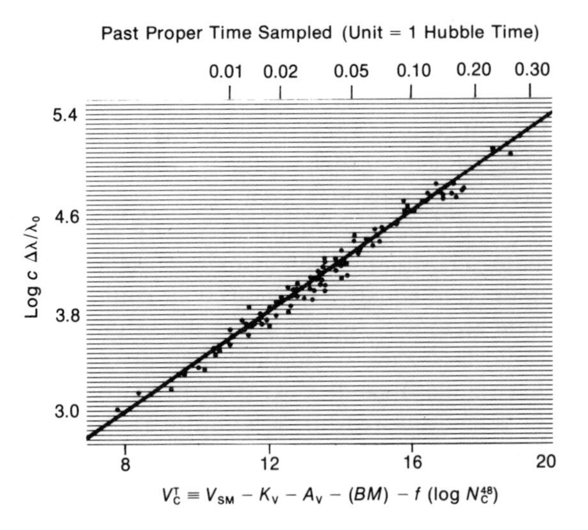

FIGURE 16-7 Magnitude-redshift diagram for different galaxies. The vertical axis plots redshifts, and the horizontal axis plots magnitudes, the same as in the right panel of Figure 16-6. K_V, A_V, (BM), and $f\,(\log N_c{}^{48})$ are all correction terms. (From A. Sandage and E. Hardy, *Astrophysical Journal*, vol. 183, p. 755, 1973, published by the University of Chicago Press. Copyright © by the American Astronomical Society. All rights reserved.)

have shifted, and convert to z = wavelength shift per wavelength when the photon left the galaxy. The magnitudes are more difficult. The symbols on the label for the horizontal axis all refer to different corrections. I left this caption on the figure not to torture you with all sorts of symbols to remember but to point out how complex the correction procedure is. For those of you who want to know what these symbols mean, here are some brief indications.

The galaxy absorbs some light, so you have to take that out, A_V. When you measure magnitudes of galaxies of different redshifts, you look at different wavelengths as the spectrum of the galaxy is redshifted. This correction and corrections for the geometry of the universe are called K_v. The last two symbols, (BM) and f $(\log N_c^{48})$, sound horrible and indeed they are. As you look backward into time you are looking at fainter and fainter galaxies, and you tend to pick out the brightest galaxies to observe, so you no longer are looking at standard candles. As we shall see later, these selection effects may be large.

For the moment, let us assume that we have made all of the corrections unerringly and blast ahead to see what happens. A graph like Figure 16-7 is not too illuminating, as the overall slope of the data to the right obscures the effects from evolution of the universe. In Figure 16-8, the same data have been replotted in the sense that deviations from the $q = 1$ line are shown on the horizontal axis. At first glance, the data seem to indicate that the universe is closed, because you can use curve-fitting theory to see which theoretical curve best fits the observations. Mathematically, you find $q = 1.13$. If you leave out the three most distant clusters, however, for which the data are not very certain, the value of q becomes 0.65, with a large uncertainty. It might seem that the universe is closed. But wait.

You must be careful not to take these accurate-sounding values for q too literally. Look again at Figure 16-8. The uncertainties in the data, as shown by the scatter of the observations around the curves, are tremendous. One nice feature of curve-fitting theory is that it allows you to estimate your possible error. Including the observations of the three most distant galaxies, it turns out that there is a 50 percent probability that q is somewhere between 0.77 and 1.46. Fine, you say; so the universe is closed. You should remember that the nice-sounding 50 percent probability means also that there is a 50 percent probability that q is *not* within that range, or that if you interpret this data literally, there is a 10 percent probability that the universe is open, even if you assume that all the corrections have been applied correctly. The data seem to favor a closed universe.

Allan Sandage of the Hale Observatories has published data of this sort for some time now. Earlier conclusions from these data that we live in a closed universe spurred the search for the missing mass, chronicled in the last section. Yet there is a crucial assumption involved in literally accepting the data. Remember, we have been assuming that the brightest galaxies in clusters of galaxies were standard candles, equal in luminosity wherever and whenever they are seen in the universe. We are looking backward in time to see changes in the redshift; perhaps there are changes in the luminosity of elliptical galaxies too? Early work indicated that these brightest galaxies, generally ellipticals, had not changed in luminosity recently, over the last two billion years or so. It was only in the 1970s that the real magnitude of the evolutionary effects was understood.

In general, you expect that galaxies would have been more luminous in the past. As a galaxy evolves, massive stars in it die, and are no longer seen. If galaxies had been more luminous in the past, they would be moved along the arrows shown in the right panel of Figure 16-6. If you ignored the effects of this evolution, the universe would seem more closed than it really is.

Beatrice Tinsley of the University of Texas calculated what the effects of galactic evolution would be, and her work indicated that they were unpleasantly large. She took a model of an elliptical galaxy, including different numbers of different types of stars. She let these stars evolve, according to the evolutionary tracks of the stellar evolutionists, and watched, through the computer, what happened to her model galaxy. She found that the evolutionary effects are as large as the effects that come from the changes in the expansion rate. It is not completely certain what these effects are, as the theory that goes into them is not completely worked out. But it seems that the results obtained from diagrams like Figure 16-8 should be viewed with caution.

The importance of these evolutionary effects is illustrated by Allan Sandage's changing views on the subject, for example. In 1971, he argued that "at the present rate of progress, perhaps five years will be required for the solution, but optimistic observers can see no unsolvable problem blocking the road," as he discussed the magnitude-redshift relations of this section.[2] The unseen, at present unsolved problem appeared by 1973, as Sandage wrote: "The true q_0 value depends on unknown evolutionary effects not discussed here."[3] This is the way in which science works. The earlier work on these diagrams was not wasted, for until we attempted to answer the riddle of the future of the universe by looking at redshifts of elliptical galaxies, no one knew whether or not it would be interesting to see what the evolutionary effects were.

Another problem with the magnitude-redshift relation has arisen. James Gunn and J. B. Oke of CalTech have shown that the selection of the galaxies to observe in making graphs like Figure 16-8 has a critical effect on the outcome of the test. You tend to pick out the brightest and thus the most luminous galaxies to observe because those are the ones

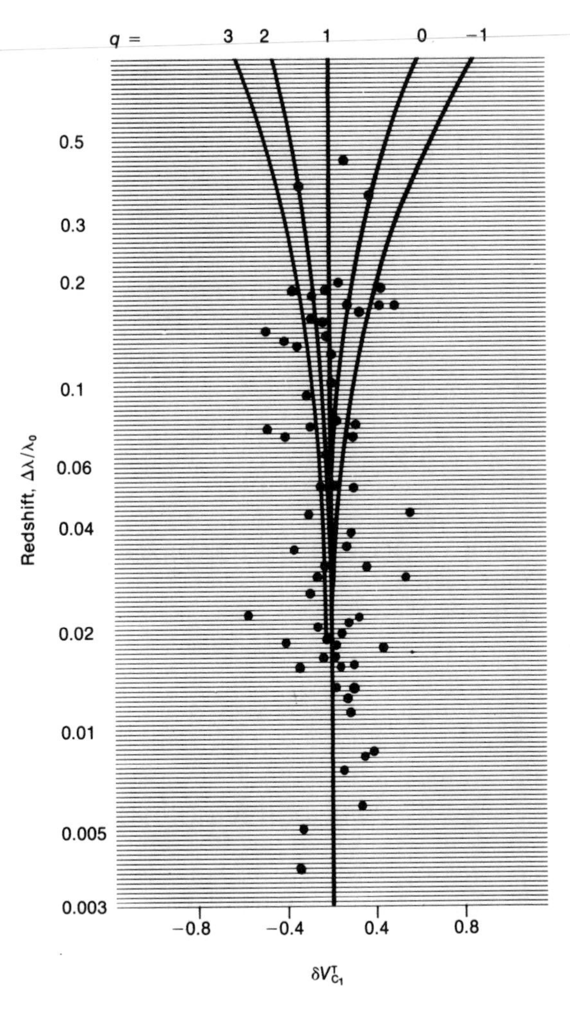

FIGURE 16-8 Figure 16-7 replotted, so that the difference $\delta V^T{}_{c_1}$ between the observations and the $q=1$ line is on the horizontal axis. (From A. Sandage and E. Hardy, *Astrophysical Journal,* vol. 183, p. 755, 1973, published by the University of Chicago Press. Copyright © by the American Astronomical Society. All rights reserved.)

you can see best. In fact, at large redshifts where the effects of q or changing expansion rate are observable, the brightest galaxies are the only ones you can see. The preliminary results of these investigators, presented at a scientific meeting in the summer of 1974, indicate that proper allowance for these selection effects should produce a value of q near zero; but their chief contribution at this date is to indicate that these selection effects are very important, and that until they are understood it is impossible to use this test to decide the future of the universe. We must seek another way to decide whether the universe is closed or open.

Deuterium

Deuterium is heavy hydrogen. Ordinary hydrogen has just one proton in its nucleus, whereas deuterium has one proton and one neutron. In 1973 and 1974, with the first discoveries of deuterium in astronomical objects, a new way of determining whether the universe is closed or open has appeared. Deuterium can serve as a probe into the mass density of the early universe, and you can then use the mass density obtained from such a study to determine whether the universe now contains enough mass to be closed.

When the early universe fused hydrogen to form helium, deuterium was the first step on the road. Most of the deuterium that was made subsequently became helium, for the universe was dense enough at that time for the deuterium to fuse with neutrons and protons, building up to two protons and two neutrons — helium. A small fraction of the deuterium managed to survive the early helium-burning epoch and survive to the present day. How much deuterium managed to survive depends on the density of the universe. Directly related to the density of the universe was the likelihood that a primeval deuterium atom would collide with other particles and become something heavier. A dense, closed universe would not have much cosmological deuterium as all the deuterium would have become helium, and a rarefied, open universe would leave the deuterium. These results are displayed quantitatively in Figure 16-9. By measuring how much deuterium is around, you can then determine the present density of the universe by reading it from a curve like Figure 16-9. Here is another way to answer the cosmological question, Is the universe closed?

Applying this idea was seriously limited by the complete absence, before 1972, of any discovery of deuterium anywhere in the universe except on the earth. In the last two years, however, discoveries of deuterium have been popping up in the literature like mushrooms, as six discoveries of deuterium in five objects have been made. Yet finding deuterium is not the same thing as measuring its cosmic abundance. The only straightforward abundance measurement was made by Princeton astronomers on the Copernicus satellite, which measured absorption lines from the interstellar medium in the star Beta Centauri. The satellite was needed for the observations because the deuterium lines lie in the ultraviolet part of the spectrum, which is absorbed by our atmosphere. The Princeton group found a deuterium abundance of 1.5×10^{-5}.

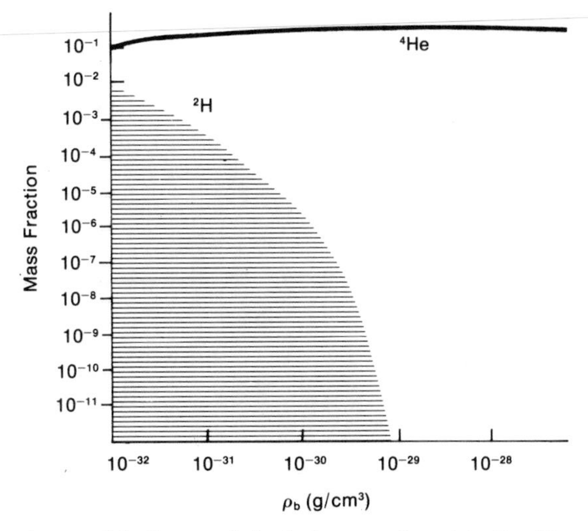

FIGURE 16-9 The abundance of helium and deuterium produced in the Big Bang versus the present density of the Universe ρ_b. The closure density is 4.7×10^{-30}, corresponding to a deuterium abundance of 10^{-6}. (Adapted from R. V. Wagoner, *Astrophysical Journal,* vol. 179, p. 349, 1973, published by the University of Chicago Press. Copyright © by the American Astronomical Society. All rights reserved.)

On the basis that this deuterium is really a relic of the Big Bang, all you have to do is go to Figure 16-9 and read the present density of the universe from the curve. Try it. It turns out to be about 10^{-30} g/cm^3, or about 20 percent of the closure density. If this deuterium is primordial, the universe is open.

But this test, like all the other tests of the cosmological question in this chapter, contains a trap (see Table 16-1). We do not know where this deuterium came from. It may have been made since the beginning of the universe, and as a result the universe could be a good bit denser than a straightforward interpretation would indicate. A closed universe would have a primordial deuterium abundance of 10^{-6}, and the deuterium abundance could be built up to the observed value of 1.5×10^{-5} by processes occurring subsequently. Theorists are searching for alternative ways of making this deuterium, and a couple of early ideas turned out not to work on closer examination. In view of the newness of this discovery and the absence of knowledge about the source of the deuterium in the universe, it is too early to tell much about the history of the universe from deuterium. Before a true test can be made, further discoveries of deuterium in more places must be effected. Future work will probably provide us with an important tool for probing the future of the universe.

TABLE 16-1 Cosmological Tests

TEST	VERDICT	QUALIFICATIONS
Measuring the Mass	Open	Maybe There Is Some Mass We Do Not See
Changing Expansion Rate	Closed (1970) ??? (1975)	Evolutionary and Selection Effects Are Superimportant
Deuterium	Open	Maybe the Deuterium Came from Some Other Source Besides the Big Bang

This chapter ends, as this book does, on a note of uncertainty. The results presented here are most uncharacteristic of one view of science as a nice, neat picture, but most characteristic of the confused current state of astronomy. At the moment, our knowledge of the future of the universe is clouded.

Determining the mass density of the universe provides a very tentative conclusion that the universe is open. The mass contained in galaxies is only 2 percent of the mass necessary to close the universe, and the extra mass present in the Coma cluster would, if real and if present in other clusters of galaxies, provide only 15 percent of the closure density. The elusive intergalactic medium could close the universe but has not been found yet. If the intergalactic medium is there, it exists in the form of an ionized gas.

Looking backward to determine the rate of change of the universal expansion was at one time thought to be a way of answering whether the universe is closed or not. At the present time, the roles of evolutionary effects and selection effects remain to be elucidated. Preliminary indications show that these effects, when included, turn a closed-universe interpretation of the data into an open one, but the principal impact of recent work on these effects is that they are large.

If the deuterium that has recently been found in interstellar space is a relic of the primeval fireball, then the universe is open, since it was sufficiently rarefied in the Big Bang to allow the deuterium to remain and not be processed further. Since it is not yet clear where the deuterium was made, the deuterium test is not yet conclusive.

Although we have the tools to determine the future of the universe, application is limited by the intervention of many other effects. We have to unravel a complex observational situation before we can decide the question, Is the universe open or closed? While the trend of the results in the last two years has been to reverse the trend of the 1960s, it is dangerous to rely on trends that may reverse again. The present trend of the observations in the direction of an open universe may not last, but then again it may.

SUMMARY OF PART THREE

If cosmology is to be a branch of science instead of a philosophical recreation, there must be some data that allow the real world to enter. In the 1920s and 1930s, the discovery of the expansion of the universe gave us our first handle on the evolution of the universe as a whole, and the logical interpretation of that expansion was the Big Bang theory. Until 1960, however, it was impossible to decide between the Big Bang and Steady State theories, because the expansion of the universe was the only piece of information that we had on its evolution. We could measure the expansion rate and determine the age of the universe, but we could not do so accurately.

Subsequent developments have greatly refined our knowledge of cosmology. The discovery of the primeval fireball, the background radiation filling the universe, which is a Big Bang relic, confirmed the Big Bang picture. The existence of a cosmic helium abundance strengthens the Big Bang theory, but the interpretation of the helium data is still not certain. If the radio sources and quasars are very distant objects and we are seeing them as they were when the universe was still young, their abundance in the early universe rules out the Steady State theory in its original, strictly interpreted form. As a result we are fairly confident about our ideas of the past history of the universe.

We are much less certain of the future. Will the universe go on expanding forever? A search for the necessary mass to slow down the expansion did not succeed in its being uncovered, but there are places that undetected mass could be hiding. Efforts to look to the past to detect changes in the expansion rate are clouded by problems of interpretation. Recent discoveries of deuterium seem to be pointing in the direction of an open universe, but it is not yet clear where this deuterium came from.

The Big Bang theory leaves one unanswered question. Who created the material that exploded as the Big Bang? For this, the astronomer has no answer.

We may be able to look back to the early seconds of the evolution of the universe, but our vision stops there. This book ends by leaving the problems of creation to the philosopher and the theologian.

OBSERVATIONAL FACT	Existence of a microwave background Everywhere we have looked so far, the universe seems to be about 30 percent helium Deuterium exists in the universe The universe is expanding
CONCRETE THEORY	Big Bang universes make primeval fireball radiation and primordial helium; the Steady State universe contains neither
INFORMED OPINION	The microwave background is the primeval fireball (near-fact) The helium is cosmological Therefore the Steady State theory is not valid (almost everyone agrees, but we could be wrong) The Hubble constant is 50 kilometers per second per megaparsec, and so the universe is less than 20 billion years old
UNANSWERED QUESTIONS	Did the deuterium come from the primeval fireball? (If it did, the universe is open) Can we overcome the uncertainties of evolutionary and selection effects and measure changes in the expansion rate of the universe? Is there enough mass to close the universe? (We haven't found it so far)
SPECULATION	What preceded the Big Bang? The cosmological singularity? Is there any connection between this singularity and black holes?

GLOSSARY

Terms marked with an asterisk (*) are discussed in more detail in the Preliminary section.

***ANGSTROM UNIT** 10^{-8} centimeter.

ANOMALY Discrepancy between theory and observation, between the model world and the real world. One that persists may lead to a crisis.

ANTIMATTER Stuff that on being brought into contact with matter causes both matter and antimatter to disappear, leaving many gamma rays. Antimatter has so far been found only in laboratories.

***APPARENT BRIGHTNESS (APPARENT MAGNITUDE)** Measures of how bright something appears in the sky.

***ASTRONOMICAL UNIT** Mean distance between the earth and the sun (1.495985×10^8 kilometers).

B STAR, B SUPERGIANT A hot star. A supergiant is an extremely large and luminous star. The letter "B" refers to the star's spectral class.

BINARY STAR (BINARY SYSTEM) A pair of stars orbiting each other.

***BLACKBODY RADIATION** Emitted by an object with no intrinsic color. It has a characteristic spectrum, or distribution of intensity in different wavelengths, which is independent of the nature of the object emitting the blackbody radiation.

BLINK MICROSCOPE Device that allows alternate viewing of two photographs of the same part of the sky. It is used to discover variable stars and moving objects like asteroids.

CEPHEID VARIABLES Stars that vary in luminosity in a regular, periodic fashion. If you know the period of a Cepheid variable, you can determine its luminosity.

COSMOLOGICAL REDSHIFT Shift in spectrum lines due to the expansion of the universe.

CRISIS Stage in a scientific revolution when an anomaly has grown important enough to cause people to question the validity of the prevailing paradigm.

DECELERATION PARAMETER Number that measures rate of change of the expansion of the universe. Sometimes called q_0. If the value is larger than 0.5, the universe is closed and will eventually stop expanding.

DECLINATION Celestial coordinate used to measure star positions, roughly the equivalent of latitude.

*DEGREE Unit of measure of angle. The sun and moon are each half a degree across in the sky.

DENSITY Amount of matter per unit volume. Not to be confused with mass, which measures amount of matter.

DISCORDANT REDSHIFT Hypothetical redshift in the spectrum lines of a quasar that arises from no known physical cause. (Chapter 12 discusses the evidence for and against the existence of discordant redshifts.)

DISTANCE MODULUS Difference between apparent and absolute magnitudes of an object. (See Chapter 14.)

DOPPLER SHIFT Wavelength change that photons make when they travel from an object moving relative to the observer. (See Chapter 5.)

*ELECTRON One of the three types of particles found in an atom. It has a negative electrical charge and is found around the nucleus of the atom.

*ELECTRON-VOLT Unit of energy, equal to 1.60207×10^{-12} erg. Atomic energy levels are typically separated by a few electron-volts.

*ERG Fundamental unit of energy. A two-gram insect crawling along at a speed of 1 centimeter per second has an energy of motion of 1 erg.

GLOBULAR CLUSTER Group of old stars, generally found in the halo of a galaxy.

HOMOGENEOUS Uniform, not lumpy.

*KELVINS Unit for measuring temperature; degrees Celsius above absolute zero.

KEPLER'S THIRD LAW Relation between period, separation, and mass of two objects orbiting each other. It is generally used to measure the mass of the orbiting objects; you can determine the characteristics of the orbit by observation.

*KILOPARSEC 1000 parsecs.

LUMINOSITY Amount of energy that an object radiates into space every second.

*MAGNITUDE Scale astronomers use to measure brightness of objects.

MAIN-SEQUENCE STAR — Star that supports itself by fusion of hydrogen in its center. The sun is a main-sequence star.

*MEGAPARSEC — 10^6 (one million) parsecs.

*MICROMETER — 10^{-6} meter.

*MICROSECOND — 10^{-6} second.

*MICROWAVE RADIATION — Radio-frequency radiation of extremely short wavelength, and infrared radiation of long wavelength, generally radiation with wavelengths roughly between 100 micrometers and 1 centimeter.

*MINUTE OF ARC — Unit of measure of angle, 1/60 of a degree.

NEUTRINO — Particle that travels at the speed of light and rarely reacts with anything.

*NEUTRON — One of the particles found in an atomic nucleus. While roughly equal in mass to the proton, it has no electrical charge.

NORMAL SCIENCE — The usual state of science, with all research based on a prevailing paradigm.

PARADIGM — A set of scientific laws that is a basis for scientific work. See Chapter 1.

*PARSEC — 3.2615 light-years, 206264.8 astronomical units, or 3.1×10^{13} km. In our part of the galaxy, stars are a parsec apart, on the average.

PHOTOGRAPHIC PLATE — Piece of glass coated with photographic emulsion. Film is celluloid coated with emulsion. Astronomers generally use plates instead of film because glass does not stretch.

*PHOTON — A light particle.

PHOTOSPHERE — Visible surface of a star. This term generally is used to describe the sun.

PLASMA — An ionized gas, composed of electrons and atoms missing one or more electrons.

POSITRON — A positively charged antielectron. When a positron and an electron collide, they annihilate each other.

PROTON — One of the particles found in the nucleus of an atom. It has positive electrical charge.

RECOMBINATION — Process in which an electron and an ion collide, stick together, and form an atom (or less highly charged ion). This term is also used to refer to the time that matter in the universe recombined, 700,000 years after the beginning.

RIGHT ASCENSION — Celestial coordinate used to measure star positions, roughly the equivalent of longitude.

SCHMIDT TELESCOPE	Type of telescope that has a wide field of view, useful in survey work.
*SECOND OF ARC	Unit of measure of angle, 1/60 minute of arc, 1/3600 degree.
SINGLE-LINED SPECTROSCOPIC BINARY	Star that orbits an invisible object. The star's motion shows up through changing Doppler shifts in the star's spectrum.
*SPECTRAL CLASSIFICATION	Scheme that allows determining the general properties of a star by spectrum examination. Spectral classes are denoted by letters — O, B, A, F, G, K, M.
*SPECTRUM	Graph or photograph of the intensity of radiation from an object as it varies with wavelength.

SUGGESTIONS FOR FURTHER READING

Items marked by asterisks (*) should be intelligible to anyone who has read this book. Other items are more technical.

PRELIMINARY

*Abell, George O. *Exploration of the Universe,* 2nd ed. Holt, Rinehart, and Winston, New York, 1969. (A basic text, with a new edition expected soon.)

*Asimov, Isaac. *The Universe: From Flat Earth to Quasar.* Avon, New York, 1966. (In paperback.)

*Hodge, Paul W. *Concepts of Contemporary Astronomy.* McGraw-Hill, New York, 1974.

*Jastrow, Robert, and Malcolm Thompson. *Astronomy: Fundamentals and Frontiers,* 2nd ed. Wiley, New York, 1974.

These are all basic texts. There are many, many others.

CHAPTER 1 Introduction: The Violent Universe

*Calder, Nigel. *Violent Universe.* Viking, New York, 1969. (Based on a BBC television series and well illustrated, this book describes the observations leading to the discoveries discussed in this book.)

*Kuhn, Thomas S. *The Structure of Scientific Revolutions.* University of Chicago Press, Chicago, 1967. (The classic exposition.)

*Synge, J. L. *Talking About Relativity.* Elsevier, New York, 1970.

*Toulmin, Stephen. *Foresight and Understanding.* Harper, New York, 1961. (He argues that prediction is not the essence of science.)

CHAPTER 2 Stellar Evolution: To the White Dwarf Stage

Chiu, Hong-Yee, and Amador Muriel, eds. *Stellar Evolution.* MIT Press, Cambridge, 1972. (Technical.)

Clayton, Donald D. *Principles of Stellar Evolution and Nucleosynthesis.* McGraw-Hill, New York, 1968. (Graduate-level text.)

*Greenstein, Jesse L. "Dying Stars." In *Frontiers in Astronomy,* ed. Owen Gingerich. Freeman, San Francisco, 1971. (This book is a collection of articles from *Scientific American* and covers a wide variety of topics in present-day astronomy. A new edition with some more recent articles is expected soon.)

*Meadows, A. J. *Stellar Evolution.* Pergamon, Oxford, 1967. (For nonspecialists, in paperback.)

Ostriker, Jeremiah P. "Recent Developments in the Theory of Degenerate Dwarfs." *Annual Review of Astronomy and Astrophysics* 9 (1971), 353–366.

Weidemann, Volker. "White Dwarfs." *Annual Review of Astronomy and Astrophysics* 6 (1968), 351–372.

CHAPTER 3 Supernovae, Neutron Stars, and Pulsars

*Hewish, Anthony. "Pulsars." In *Frontiers in Astronomy,* ed. Owen Gingerich. Freeman, San Francisco, 1971. (*Scientific American* level; see above.)

Hewish, Anthony. "Pulsars." *Annual Review of Astronomy and Astrophysics* 9 (1971), 353–366.

*Hewish, Anthony. "Pulsars and High Density Physics." Nobel Prize Lecture, 1974. *Science* 188 (June 13, 1975), 1079–1083.

*Lovell, A. C. B. *Out of the Zenith.* Harper, New York, 1973. (Historical.)

Ruderman, M. "Pulsars, Structure and Dynamics." *Annual Review of Astronomy and Astrophysics* 10 (1972), 427–476.

Shklovsky, I. S. *Supernovae.* Wiley, London, 1968.

*Verschuur, Gerrit L. *The Invisible Universe.* Springer-Verlag, New York, 1974. (This book describes radio astronomy at the nonspecialist level.)

Wheeler, J. Craig. "After the Supernova, What?" *American Scientist* 61 (1973), 42–51.

CHAPTER 4 Journey into a Black Hole

*Kaufmann, William J., III. *Relativity and Cosmology.* Harper, New York, 1973. (Includes special and general relativity, black holes, and speculative ideas, written for the nonspecialist.)

Misner, C. W., K. S. Thorne, and J. A. Wheeler. *Gravitation.* Freeman, San Francisco, 1973. (A superior text for a senior or graduate-level course on Einstein's theory of gravity. This book, and Kip Thorne's course based on it, has both inspired and enlightened me.)

*Penrose, Roger. "Black Holes." *Scientific American* (May 1972), 38–46.

*Penrose, Roger. "Black Holes." In *Cosmology Now,* ed. Laurie John. BBC Publications, London, 1974. (The book is a good reference on black holes and especially on cosmology.)

*Thorne, Kip S. "Gravitational Collapse." *Scientific American* (November 1967), 88–98. (Excellent summary.)

Will, Clifford M. "Einstein on the Firing Line." *Physics Today* (October 1972), 23–31.

CHAPTER 5 The Search for Black Holes

*Gursky, Herbert, and Edward P. J. van den Heuvel. "X-Ray

Emitting Double Stars." *Scientific American* (March 1975), 24–35.

*Jones, Christine, William Forman, and William Liller. "X-Ray Sources and Their Optical Counterparts." A three-part series in *Sky and Telescope* 48 (November 1974), 289–291; 48 (December 1974), 372–375; 49 (January 1975), 10–13. Part 1 of this series is most directly relevant to this section.

Thorne, Kip S. "The Search for Black Holes." *Scientific American* (December 1974), 32–43.

Many of the fascinating details of the stories of Cygnus X-1 and Epsilon Aurigae have been confined to the technical literature. You nonspecialists should not be put off by the fact that the papers below are in technical journals; if you are willing to accept that you will not understand all of each paper you can learn a good bit from reading them. I list some of the highlights:

Epsilon Aurigae

Alaistair G. W. Cameron, "Evidence for a Collapsar in the Binary System Epsilon Aur," *Nature* 229 (1971), 178–179, originally proposed the black hole idea. R. Stothers, "Collapsars, Infrared Disks, and Invisible Secondaries of Massive Binary Systems," *Nature* 229 (1971), 180–183, agreed. For counterarguments, see P. Demarque and S. C. Morris, "Is There a Black Hole in Epsilon Aurigae?", *Nature* 230 (1971), 516–517. Recent models of Epsilon Aurigae are R. E. Wilson, "A Model of Epsilon Aurigae," *Astrophysical Journal* 170 (1971), 529–539, and Su-Shu Huang, "Interpretation of Epsilon Aurigae, II," *Astrophysical Journal* 187 (1974), 87–92. Huang argues that there is evidence for condensations in the disk that may indicate the formation of a planetary system.

Cygnus X-1

While the two articles cited above tell the current situation, the history can be found in the following list of selected papers. The original identification of Cygnus X-1 with the blue star HDE 226868 was announced in several papers in the August 15, 1971 issue of *Astrophysical Journal Letters* (which in this paragraph shall be called *ApJL* since it comes up so often): S. Miyamoto, M. Fujii, M. Matsuoka, J. Nishimura, M. Oda, Y. Ogawara, S. Ohta, and M. Wada, "Measurement of the Location of the X-Ray Source Cygnus X-1," *ApJL* 168 (1971), L11–L15; A. Toor, R. Price, F. Seward, and J. Scudder, "X-Ray Source Positions for Cygnus X-1, Cygnus X-2, and Cygnus X-3," *ApJL* 168 (1971), L15–L16; S. Rappaport, W. Zaumen, and R. Doxsey, "On the Location of Cygnus X-1," *ApJL* 168 (1971), L17-L20; and R. M. Hjellming and C. M. Wade, "Radio Emission from X-Ray Sources," *ApJL* 168 (1971), L21–L24.

About a year later, the February 1, 1973 issue of the same journal contained the results of the 1972 observing season in papers by N. Walborn; H. E. Smith, B. Margon, and P. S. Conti; and R. Brucato and J. Kristian, *ApJL* 179 (1973),

123–134. The HZ 22 episode is found in J. L. Greenstein, "A Highly Evolved Low-Mass Binary: HZ 22," *Astronomy and Astrophysics* 23 (1973), 1–7; V. Trimble, "Low-Mass B Stars with Low Surface Gravity," *Astronomy and Astrophysics* 23 (1973), 281–283. The connection between HZ 22 and Cygnus X-1 was announced by V. Trimble, J. Weber, and W. Rose in "A Low-Mass Primary for Cygnus X-1?" *Monthly Notices of the Royal Astronomical Society* 162 (1973), 1p–3p. This idea bit the dust when it was shown that Cygnus X-1 was a very distant star by B. Margon, S. Bowyer, and R. P. S. Stone, "On the Distance to Cygnus X-1," *ApJL* 185 (1973), 113–116; J. Bregman, D. Butler, E. Kemper, A. Koski, R. P. Kraft, and R. P. S. Stone, "On the Distance to Cygnus X-1 (HDE 226868)," *ApJL* 185 (1973), L117–L120.

Alternative models for Cygnus X-1 are J. N. Bahcall, M. N. Rosenbluth, R. M. Kulsrud, "Model for X-Ray Sources Based on Magnetic Field Twisting," *Nature Physical Science* 243 (1973), 27–28; K. Brecher and P. Morrison, "Rapidly Rotating Degenerate Dwarfs as X-Ray Sources in Binaries," *ApJL* 180 (1973), L107–L112; and J. N. Bahcall, F. J. Dyson, J. I. Katz, and B. Paczynski, "Multiple Star Systems and X-Ray Sources," *ApJL* 189 (1974), L77–L78; and A. C. Fabian, J. E. Pringle, and J. A. J. Whelan, "Is Cyg X-1 a Neutron Star?" *Nature* 247 (1974), 351, argue for a triple system. But see G. R. Blumenthal and W. Tucker, "Compact X-Ray Sources," *Annual Review of Astronomy and Astrophysics* 12 (1974), 23–46, and H. L. Shipman, "The Implausible History of Triple Star Models for Cygnus X-1: Evidence for a Black Hole," *Astrophysical Letters* 16 (1975) 9–12.

I know it's a long list. But as long as you don't let yourself, as a nonspecialist reader, be intimidated by the equations, a look at some of these papers will make you realize what an uncertain business science is.

CHAPTER 6 Frontiers and Fringes

Hawking, S. W., and G. F. R. Ellis. *The Large Scale Structure of Space-Time.* Cambridge University Press, Cambridge, U.K., 1973.

Most of the references listed in Chapter 4 deal with this material also, especially Kaufmann's book. Black holes are the subject of discussion at the yearly Texas conferences on high-energy astrophysics, and proceedings are published. The last one, the sixth, is published in *Annals of the New York Academy of Sciences* 224 (1973).

CHAPTER 7 Galaxies Near and Far

Burbidge, E. M., and G. R. Burbidge. *Quasi Stellar Objects.* Freeman, San Francisco, 1967. (At the intermediate level.)

Hodge, P. W. *Galaxies and Cosmology.* McGraw-Hill, New York, 1966.

*National Academy of Sciences Astronomy Survey Committee. *Astronomy and Astrophysics for the 1970's.* National Academy of Sciences, Washington, 1972. (This superb report, prepared by a panel of astronomers under the leadership of Jesse L. Greenstein, looks ahead at the future of astronomy.)

O'Connell, D. J. K. *Nuclei of Galaxies.* Elsevier, New York, 1971. (This is the proceedings of a study week held at the Vatican on the subjects of Part Two, and is a good review and general reference.)

When Volume 9 of *Stars and Stellar Systems* (University of Chicago Press) appears, it will be a standard reference on galaxies.

CHAPTER 8 Radio Waves from Quasars

*Hey, J. S. *The Evolution of Radio Astronomy.* Neale Watson, New York, 1973.

*Lovell, A. C. B. *The Story of Jodrell Bank.* Harper, New York, 1968.

Two books listed above under Chapter 3 are also helpful here: Verschuur, *The Invisible Universe,* and Lovell, *Out of the Zenith.*

CHAPTER 9 An Optical Astronomer's View of Quasars

Middlehurst, B. M., and L. H. Aller, eds. *Stars and Stellar Systems,* vol. 7. University of Chicago Press, Chicago, 1967. (A discussion at the advanced level of how emission lines are produced.)

Osterbrock, D. "Physical Conditions in the Active Nuclei of Galaxies and Quasi-Stellar Objects." In *Nuclei of Galaxies,* ed. D. J. K. O'Connell. Elsevier, New York, 1971.

CHAPTER 10 Active Galaxies

Burbidge, G. R. "Nuclei of Galaxies." *Annual Review of Astronomy and Astrophysics* 8 (1970), 369–460. (Fairly advanced, but comprehensive.)

*Jones, Christine, William Forman, and William Liller. "Optical Counterparts of X-Ray Sources — III." *Sky and Telescope* 49 (January 1975), 10–13.

*Kellermann, K. I. "Extragalactic Radio Sources." *Physics Today* (October 1973), 38–47.

*Metz, William D. "Double Radio Sources: Energetic Evidence that Galaxies Remember." *Science* 188 (June 27, 1975), 1289–1292.

*Sanders, R. H., and G. T. Wrixon. "The Center of the Galaxy." *Scientific American* (April 1974) 67–78.

*Weymann, R. J. "Seyfert Galaxies." In *Frontiers in Astronomy,* ed. Owen Gingerich. Freeman, San Francisco, 1971.

CHAPTER 11 The Energy Source of Quasars

*Pacini, Franco, and Martin Rees. "Rotation in High-Energy Astrophysics." *Scientific American* (February 1973), 98–105.

*Schmidt, Maarten, and Francis Bello. "The Evolution of Quasars." *Scientific American* (May 1971) 54–69.

The second part of D. J. K. O'Connell's *Nuclei of Galaxies* (see above under Chapter 7) contains many discussions of quasar theories.

CHAPTER 12 Alternative Interpretations of the Redshift

*Arp, Halton C. "Observational Paradoxes in Extragalactic Astronomy." *Science* 174 (1971), 1189–1199. (A very intelligible summary of the case for discordant redshifts in 1971. You should read it for a forceful presentation of Arp's point of view, which is different from mine.)

*Field, George B., Halton Arp, and John N. Bahcall. *The Redshift Controversy.* Benjamin, Reading, Mass., 1973. (A gold mine. Here you have a statement of the cases for and against discordant redshifts, plus reprints of the seminal papers in the controversy. The asterisk is not a mistake. Nonspecialists who do not let themselves be intimidated by equations can understand this.)

CHAPTER 13 Life Cycle of the Universe: A Model

*Gamow, George. *The Creation of the Universe.* Mentor, New York, 1952. (Old but still good. The numbers are no longer correct, but the general picture is valid.)

Harrison, E. R. "Standard Model of the Early Universe." *Annual Review of Astronomy and Astrophysics* 11 (1973), 155–186. (A comprehensive review at the advanced undergraduate level. He sets the historical record straight, among other things.)

*John, Laurie, ed. *Cosmology Now.* BBC Publications, London, 1974. (A good review of cosmology, based on a television series.)

Peebles, P. James E. *Physical Cosmology.* Princeton University Press, Princeton, 1971. (A good text, the basis of much of Part Three.)

*Schramm, David N. "The Age of the Elements." *Scientific American* (January 1974), 69–77.

Sciama, D. W. *Modern Cosmology.* Cambridge University Press, Cambridge, U.K., 1971. (Good review, less mathematical than Peebles.)

*Singh, Jagjit. *Great Ideas and Theories of Modern Cosmology,* 2nd ed. Dover, New York, 1970. (Sometimes the going is a bit rough, but this is a good overview of cosmology. Singh devotes more attention to unconventional cosmologies than I do.)

Many of these references also cover the topics of Chapters 14, 15, and 16.

CHAPTER 14 The Cosmic Time Scale

*Iben, I., Jr. "Globular Cluster Stars." *Scientific American* (July 1970), 26–39.

*Kraft, R. P. "Pulsating Stars and Cosmic Distances." In *Frontiers in Astronomy,* ed. Owen Gingerich, Freeman, San Francisco, 1971.

Sandage, Allan. "Distances to Galaxies, the Hubble Constant, and the Edge of the World." *Quarterly Journal of the Royal Astronomical Society* 13 (1972), 282–296.

A series of *Astrophysical Journal* articles on Sandage's latest results on the Hubble constant has been published in the *Astrophysical Journal* in 1974 and 1975, under the names of Sandage and G. A. Tammann. The references for Chapter 13 are also applicable here.

CHAPTER 15 Relics of the Big Bang

*Burbidge, G. R. "Was There Really a Big Bang?" In W. C. Saslaw and K. C. Jacobs, *The Emerging Universe.* University of Virginia Press, Charlottesville, 1972.

Danziger, I. J. "The Cosmic Abundance of Helium." *Annual Review of Astronomy and Astrophysics* 8 (1970), 161–178.

*Field, G. B. "Big Bang Cosmology: The Evolution of the Universe." In W. C. Saslaw and K. C. Jacobs, *The Emerging Universe.* University of Virginia Press, Charlottesville, 1972.

*Pasachoff, Jay M., and William A. Fowler. "Deuterium in the Universe." *Scientific American* (May 1974), 108–118.

*Peebles, P. J. E., and Wilkinson, D. T. "The Primeval Fireball." In *Frontiers in Astronomy,* ed. Owen Gingerich. Freeman, San Francisco, 1971.

*Ryle, M. "Radio Telescopes of Large Resolving Power." Nobel Prize Lecture, 1974. *Science* 188 (June 13, 1975), 1071–1078.

Thaddeus, P. "The Short Wavelength Spectrum of the Microwave Background." *Annual Review of Astronomy and Astrophysics* 10 (1972), 305–334.

CHAPTER 16 The Future of the Universe

Most of the references listed above are good here. In particular, Chapter 4 of Peebles, *Physical Cosmology* (see above, Chapter 13) and Pasachoff and Fowler, "Deuterium" (see above, Chapter 15) deal directly with these issues. Allan Sandage, in "Cosmology: A Search for Two Numbers," *Physics Today* (February 1970), 34–41, writes for the nonspecialist, but he was somewhat more optimistic then than he is today.

NOTES

CHAPTER 3

1. I. S. Shklovsky, *Supernovae,* London, Wiley, 1968, p. 51 (translation by Duyvendak).
2. Ibid., p. 44.
3. W. Baade and F. Zwicky, "Supernovae and Cosmic Rays," *Physical Review* 45 (1934), p. 128.
4. J. Craig Wheeler, "After the Supernova, What?" *American Scientist* 61 (1973), pp. 42–51.

CHAPTER 4

1. The times in Table 4-1 are given for a particle falling from infinity, taking as time 0 the instant that the particle is 1 a.u. away from the ten-solar-mass hole.

CHAPTER 5

1. *Nature* 243 (1973), p. 114.

CHAPTER 7

1. Various names for the Milky Way are discussed in the source for this quotation: R. H. Allen, *Star Names: Their Lore and Meaning,* New York, Dover, 1963 (originally published in 1899), p. 476.

CHAPTER 8

1. All of the information on 4C 39.25, including the quotation, comes from D. B. Shaffer's Ph.D. thesis, California Institute of Technology, 1973, p. 44. I thank him for discussions on VLB interferometry.

CHAPTER 9

1. G. R. Burbidge, T. W. Jones, and S. L. O'Dell, "Physics of Compact Nonthermal Sources. III. Energetic Considerations," *Astrophysical Journal* 193 (1974), pp. 43–54.
2. G. MacAlpine, "Photoionization Models for the Emission Line Region of Quasi Stellar and Related Objects," *Astrophysical Journal* 175 (1972), pp. 11–30; G. A. Shields, J. B. Oke, and W. L. W. Sargent, "The Optical Spectrum of the Seyfert Galaxy 3C 120," *Astrophysical Journal* 176 (1972), pp. 75–90; K. Davidson, "Photoionization and the Emission Line Spectra of Quasi Stellar Objects," *Astrophysical Journal* 171 (1972), pp. 213–232.
3. Data from F. J. Low, "Observations of 3C 273 and 3C 279 at a Wavelength of 1 mm," *Astrophysical Journal* 142 (1965), pp. 1287–1288; D. E. Kleinmann and F. J. Low, "Observations of Infrared Galaxies," *Astrophysical Journal Letters* 159 (1970), pp. L165–L172; B. Margon, S. Bowyer, and M. Lampton, "Limits on Intergalactic Helium from the 3C 273 X-Ray

Spectrum," *Astrophysical Journal* 174 (1972), pp. 471–475. I thank Dr. M. Simon for discussions on this figure.

CHAPTER 10

1. T. Matthews, W. W. Morgan, and M. Schmidt, "A Discussion of Galaxies Identified with Radio Sources," *Astrophysical Journal* 140 (1964), pp. 35–49.
2. A. Sandage and N. Visvanathan, "Linear Polarization of the $H\alpha$ Emission Line in the Halo of M 82 and the Radiation Mechanism of the Filaments," *Astrophysical Journal* 176 (1972), p. 51.
3. W. L. W. Sargent, in D. J. K. O'Connell, ed., *Nuclei of Galaxies,* Elsevier, New York, 1971, p. 82.
4. As some of these data are quite recent or uncertain or both, some readers might want to know the sources. Entries are coded 1=spiral; 2=M 82; 3=M 87; 4=N-galaxy; 5=radio galaxy; 6=3C 120; 7="average" quasar; 8=3C 273. Values for spirals are based on the Milky Way and M 31; for N-galaxies on NGC 1275, 1068, and 4151; radio galaxies on Cygnus A and Centaurus A. The division between infrared and radio is placed at 3 mm. Symbols *o*=optical, *x*=x-ray, *i*=infrared, *r*=radio. *ApJ=Astrophysical Journal; ApJL=ApJ Letters.* T. A. Matthews, W. W. Morgan, and M. Schmidt, in A. Robinson, ed., *Quasi-Stellar Sources and Gravitational Collapse,* University of Chicago Press, Chicago, 1964, 1*r,* 2*r,* 3*r.* D. Gezari, R. Joyce, and M. Simon, "Observations of the Galactic Nucleus at 350 Microns," *ApJL* 179 (1973), pp. L67–L70, 1*i.* D. A. Harper and F. J. Low, "Far-Infrared Observations of Galactic Nuclei," *ApJL* 182 (1973), pp. L89–L93, 1*i,* 2*i.* W. Hoffman, C. L. Frederick, R. J. Emery, "100-micron Map of the Galactic Center Region," *ApJL* 164 (1971), pp. L23–L28, 1*i.* P. W. Hodge, *Galaxies and Cosmology,* McGraw-Hill, New York, 1966, 1*o,* 3*r,* 5*r,* 5*o.* M. S. Longair and R. A. Sunyaev, "The Universal Electromagnetic Background Radiation," *Soviet Physics Uspekhi* 14 (1972), pp. 569–599, 1*x,* 3*r,* 4*r,* 5*o.* E. Kellogg, H. Gursky, H. Tananbaum, R. Giacconi, K. Pounds, "The Extended X-Ray Source in M 87," *ApJL* 174 (1972), pp. L65–L69, 1*x,* 3*x.*

C. R. Lynds and A. Sandage, "Evidence for an Explosion in the Center of the Galaxy M 82," *ApJ* 137 (1963), 1005, 2*r.* R. Joyce, D. Gezari, M. Simon, "345-Micron Ground-Based Observations of M 17, M 82, and Venus," *ApJL* 171 (1972), pp. L67–L70, 2*i.* D. Kleinmann and F. J. Low, "Observations of Infrared Galaxies," *ApJL* 159 (1970), pp. L165–L172, a classic paper, 3*i,* 4*i,* 8*i.* G. R. Burbidge, "Nuclei of Galaxies," *Annual Review of Astronomy and Astrophysics* 8 (1970), pp. 369–460, 4*r,* 6*r.* F. J. Low, in D. J. K. O'Connell, ed., *Nuclei of Galaxies,* Elsevier, New York, 1971, 4*i,* 6*i,* 7*i,* 8*i.* K. Kellermann, B. G. Clark, D. L. Jauncey, J. J. Broderick, D. B. Shaffer, M. H. Cohen, and A. E. Niell, "Observations of Further Outbursts in the Radio Galaxy 3C 120," *ApJL* 183 (1973), pp. 51–56, 6*r.* W. Tucker, E. Kellogg, H. Gursky, H. Tananbaum, "X-Ray Observations of NGC 5128 (Centaurus A) from UHURU," *ApJ* 180 (1973), p. 715, 5*x.* A. Sandage, in

Nuclei of Galaxies, p. 271, 4*o,* 6*o,* 7*o.* E. M. Burbidge and G. R. Burbidge, *Quasi Stellar Objects,* Freeman, San Francisco, 1967, 7*r,* 7*o.* J. Silk, "Diffuse X and Gamma Radiation," *Annual Review of Astronomy and Astrophysics* 11 (1973), pp. 269–308, 8x. D. L. Kleinmann and E. L. Wright, "10-Micron Observations of Southern Hemisphere Galaxies," *ApJL* 191 (1974), pp. L19–L20, 5i.

CHAPTER 11

1. D. Lynden-Bell, in O'Connell, *Nuclei of Galaxies,* p. 537.

CHAPTER 12

1. This phrase is due to Lawrence H. Auer.
2. G. R. Burbidge, E. M. Burbidge, P. M. Solomon, and P. A. Strittmatter, "Apparent Associations between Bright Galaxies and Quasi-Stellar Objects," *Astrophysical Journal* 170 (1971), pp. 233–240.
3. J. N. Bahcall, in G. B. Field, H. C. Arp, and J. N. Bahcall, *The Redshift Controversy,* Benjamin, Reading, Mass., 1973, pp. 88–89, 190–194.
4. H. C. Arp, "Observational Paradoxes in Extragalactic Astronomy," *Science* 174 (1971), pp. 1189–1199.
5. F. Hoyle, in G. Field et al., *The Redshift Controversy,* pp. 307–308.
6. Ibid.
7. J. Bahcall, in G. Field et al., *The Redshift Controversy,* p. 86.
8. G. Burbidge, "The Problem of the Redshifts," *Nature Physical Science* 246 (1973), pp. 17–25.

CHAPTER 13

1. Quoted by G. Gamow, *My World Line,* Viking, New York, 1970, p. 44.

CHAPTER 14

1. D. N. Schramm, "The Age of the Elements," *Scientific American* (January 1974), pp. 69–77.

CHAPTER 16

1. D. W. Sciama, *Modern Cosmology,* Cambridge University Press, Cambridge, U.K., 1971, Chapter 8.
2. A. Sandage, in O'Connell, *Nuclei of Galaxies,* p. 620.
3. A. Sandage and E. Hardy, "The Redshift-Distance Relation. VII. Absolute Magnitude of the First Three Ranked Cluster Galaxies as Functions of Cluster Richness and Bautz-Morgan Cluster Type: The Effect of q_0," *Astrophysical Journal* 183 (1973), p. 350.

INDEX